The Chinese Air Force's First Steps Toward Becoming an Expeditionary Air Force

Cristina L. Garafola, Timothy R. Heath

Prepared for the United States Air Force

For more information on this publication, visit www.rand.org/t/RR2056

Library of Congress Cataloging-in-Publication Data is available for this publication.
ISBN: 978-0-8330-9866-5

Published by the RAND Corporation, Santa Monica, Calif.

Preface

This report is based on RAND Project AIR FORCE Strategy and Doctrine Program research that was presented at the second China Aerospace Studies Institute conference, sponsored by Headquarters, U.S. Air Force. It took place on May 2, 2016, at the RAND Corporation's Washington office in Arlington, Va. Experts on airpower, military operations, and Chinese military modernization participated in the conference and provided valuable feedback to the report's authors. The four resulting reports assess notable developments and implications of China's emerging aerospace expeditionary capabilities. As China's economic, diplomatic, and security interests continue to expand, the People's Liberation Army (PLA) and, in particular, its aerospace forces (including its air force, naval aviation, and space capabilities), will require more robust expeditionary capabilities on par with China's expanding global footprint. In addition to traditional security concerns such as Taiwan and maritime territorial disputes, such issues as countering terrorism, humanitarian assistance/disaster relief, and sea-lane protection have now become factors in the PLA's training, doctrine, and modernization efforts. In addition, command of space, to include the military use of outer space, is of increasing interest to the PLA as it seeks to develop new capabilities and operating concepts to support its growing range of military missions. This report focuses on the PLA Air Force's initial steps toward becoming an expeditionary air force, a development that will have important implications for the reach of China's military and its ability to protect China's emerging overseas interests.

RAND Project AIR FORCE

RAND Project AIR FORCE (PAF), a division of the RAND Corporation, is the U.S. Air Force's federally funded research and development center for studies and analyses. PAF provides the Air Force with independent analyses of policy alternatives affecting the development, employment, combat readiness, and support of current and future air, space, and cyber forces. Research is conducted in four programs: Force Modernization and Employment; Manpower, Personnel, and Training; Resource Management; and Strategy and Doctrine. The research reported here was prepared under contract FA7014-16-D-1000.

Additional information about PAF is available on our website: www.rand.org/paf

This report documents work originally shared with the U.S. Air Force in May 2016. The draft report, issued on March 1, 2017, was reviewed by formal peer reviewers and U.S. Air Force subject-matter experts.

Contents

Figure and Tables

Figure

Tables

Summary

In recent years, the Chinese Communist Party (CCP) has directed the People's Liberation Army (PLA) to develop capabilities to protect China's interests abroad—both regionally and globally. The PLA Air Force (PLAAF) has made incremental progress in its ability to carry out overseas operations, including organizing and deploying a long-distance strategic airlift unit capable of carrying out various nonwar missions around Asia and as far as Africa.

The PLAAF has focused heavily on developing a small number of elite units to carry out these high-profile overseas missions. The most important of these units is the Il-76-equipped 39th Regiment of the 13th Transport Division. Fighter and other aircraft have built on the experience gained from domestic long-distance deployments to take part in multilateral exercises, competitions, and demonstrations in other countries. A diverse array of combat aircraft and units has appeared in multilateral exercises, but these tend to be selected from China's most advanced platforms and elite pilots.

Through these deployments, small numbers of Chinese aircrews and technicians are learning to navigate abroad, manage issues of diplomatic access, and operate with greater autonomy. As the PLAAF gains experience through these activities, it is updating aspects of its approach to expeditionary operations to include deployed communications, logistics, and maintenance. The acquisition of larger, more capable transport planes (such as the Y-20), more experience in operating dissimilar aircraft, and greater access to foreign airfields will enable the PLAAF to better carry out its nonwar missions in Asia and around the world. Moreover, greater confidence in operating abroad will position the PLAAF to carry out a broader array of missions than it has hitherto performed. While the PLAAF's expeditionary deployments to date remain small and limited by Western standards, an increasing need to safeguard Chinese interests abroad suggests that the development of expeditionary capabilities will remain a priority for the PLAAF for years to come.

Acknowledgments

The authors would like to acknowledge the support of RAND's Project AIR FORCE; Headquarters, U.S. Air Force; and Air University and Headquarters, Pacific Air Forces, for their role in establishing the China Aerospace Studies Institute. We would also like to thank the following people for their insights and invaluable feedback throughout this process: Carl Rehberg and the staff in the Strategy Division, Deputy Chief of Staff, Strategic Plans and Requirements, Headquarters, U.S. Air Force; Lt Col Benjamin Carroll; Capt Matthew Sullivan; and our reviewers, Christopher Twomey and Cortez Cooper.

Abbreviations

ADMM	ASEAN Defense Ministers' Meeting
ASEAN	Association of Southeast Asian Nations
CCP	Communist Party of China
CMC	Central Military Commission
HA/DR	humanitarian assistance/disaster relief
Il-76	*Ilyushin*-76 (transport aircraft)
JH	*Jianhong* (fighter-bomber)
KJ-200	*Kongjing*-200 (Airborne Warning–200)
MRAF	Military Region Air Force
NEO	noncombatant evacuation operation
PLA	People's Liberation Army
PLAA	People's Liberation Army Army
PLAAF	People's Liberation Army Air Force
PLAN	People's Liberation Army Navy
POL	petroleum, oil, and lubricants
SCO	Shanghai Cooperation Organization
SMS	*The Science of Military Strategy*
TCAF	Theater Command Air Force
USAF	U.S. Air Force
VIP	Very Important Personnel

1. Introduction

In recent years, the People's Liberation Army (PLA) Air Force (PLAAF) has begun to develop nascent expeditionary capabilities. U.S. Joint Publication 1-2 defines an "expeditionary force" as an "armed force organized to achieve a specific objective in a foreign country."[1] Reflecting political sensitivities regarding the operation of military forces abroad and its own limited experience, Chinese authorities do not share this definition. However, the PLA does recognize the importance of operations abroad. For example, the 2013 edition of *The Science of Military Strategy* (SMS) discusses requirements for conveying forces abroad in order to carry out the PLAAF's core missions.[2] The 2011 edition of the *PLA Dictionary of Military Terminology* defined "strategic [power] delivery" ["战略投送"] as "the comprehensive use of all types of military transportation and units to deliver military power to a combat or crisis situation for the purposes of achieving a strategic objective." Reflecting the inherently strategic nature of such activities, the *PLA Dictionary* notes that strategic delivery activities are "generally organized by a supreme command" ["统帅部"].[3]

The Chinese definition of *strategic delivery* differs significantly from the U.S. concept of *expeditionary force* in three notable ways. First, the Chinese concept envisions strategic delivery as inherently political-strategic in nature. The operation of military forces abroad raises questions about long-standing Chinese political principles such as noninterference in another country's internal affairs, which impact decisions to develop and use overseas bases. The strategic nature of these missions can also be seen in China's focus on national-level objectives, such as the protection of overseas interests and the demonstration of the country's capabilities as a great power. Second, the Chinese concept of strategic delivery encompasses both domestic and international activities. This reality reflects both the PLAAF's limited experience abroad and the challenges of domestic operations across a large and geographically diverse country. Third, the Chinese term focuses primarily, but not exclusively, on the transportation of military forces to deal with a conflict or crisis, whereas the U.S. definition of "expeditionary force" reflects the breadth of its overseas presence in embracing a much broader array of peacetime-, crisis-, and conflict-related operations and activities. Because we seek to focus on the activities of PLAAF aircraft abroad, we will eschew the Chinese term and instead use the term *expeditionary* to refer

---

[1] U.S. Department of Defense, *Dictionary of Military and Associated Terms Joint Publication 1-2*, July 2017, p. 85.

[2] For more information on the PLAAF as it is featured in SMS 2013, see Cristina L. Garafola, "The Evolution of the PLA Air Force's Mission, Roles and Requirements," in Joe McReynolds, ed., *China's Evolving Military Strategy*, Washington, D.C.: The Jamestown Foundation, 2016, pp. 75–100.

[3] People's Liberation Army, *People's Liberation Army Military Terminology* 《中国人民解放军军语》, Beijing: Military Science Press, 2011, p. 58.

to the PLAAF's *employment of military aircraft in international operations and activities to achieve a specific purpose*. However, we acknowledge the Chinese practice of regarding the capabilities required for international operations and activities as closely linked to requirements to operate across the span of Chinese territory. Thus, our use of *expeditionary* does not exclude domestic PLAAF training, activities, and operations that have clear potential application for overseas deployments.

Mirroring trends in the activities of the PLA Navy (PLAN) and the Army (PLAA), the PLAAF has increased its international presence. However, to date, all of the activity has consisted of nonwar missions, and a substantial portion of PLAAF operations abroad have consisted of humanitarian assistance/disaster relief (HA/DR), a noncombatant evacuation operation (NEO), and personnel recovery missions involving transport aircraft. The PLAAF has organized and deployed a long-distance strategic airlift unit capable of carrying out various nonwar missions around Asia and as far as Africa. Fighter and other aircraft have built on domestic long-distance deployments to take part in multilateral exercises, competitions, and demonstrations in other countries. PLAAF aircraft have also evacuated Chinese nationals in Africa. As the PLAAF gains experience through these activities, it is updating aspects of its approach to navigation, communications, training, logistics, and maintenance. While the initial expeditionary deployments remain small and limited by Western standards, an increasing need to safeguard an expanding array of economic and strategic interests abroad suggests that the development of expeditionary capabilities will remain a priority for the PLAAF for years to come.

PLA Requirements for Expeditionary Air Forces

In late 2004, Communist Party of China (CCP) General Secretary and Chairman of the Central Military Commission (CMC) Hu Jintao announced the adoption of the "historic missions for the PLA in the new stage of the new century."[4] Reflecting security requirements to support the nation's sustained development as one of the world's economies and as an increasingly powerful country, these missions consisted of four strategic responsibilities:

- safeguarding the CCP's position as a governing party
- safeguarding the country's important period of strategic opportunity
- safeguarding China's national interests
- playing a role in bringing about world peace and common development.[5]

The third and fourth missions, in particular, raised new requirements for expeditionary capabilities. Safeguarding the country's "national interests" required the PLA to develop the

[4] Daniel Hartnett, "The New Historic Missions: Reflections on Hu Jintao's Military Legacy," in Roy Kamphausen, David Lai, and Travis Tanner, eds., *Assessing the PLA in the Hu Jintao Era*, Carlisle, Pa.: Army War College Press, 2014, p. 2.

[5] Hartnett, 2014, pp. 31–80.

capability to protect the country's economic and strategic interests, many of which span international boundaries. These national interests included vulnerable sea lines of communication, energy and natural resources, and citizens working and traveling abroad. In directing the PLA to play a role in "safeguarding world peace promoting common development," national leaders sought to leverage all elements of national power, including the military, to shape a more-favorable international security environment.[6] This directive implied the PLA should play a larger role in advancing Chinese influence and promoting the global stability and peace needed to further the nation's development.

The adoption of the historic missions thus directed the PLA to expand its operations and activities abroad to protect a growing array of national interests overseas and to shape a favorable security environment. Each of the services has accordingly developed visions to guide the modernization of its forces. In the mid-2000s, the PLAAF outlined the idea of a "strategic air force" as a guiding vision for the PLAAF's construction and operations.

Following the party's endorsement of the historic missions concept at the 17th Party Congress in 2007, PLAAF leaders stepped up efforts to develop an air force capable of carrying out expeditionary duties. In early 2008, an article in the PLAAF's official newspaper *Air Force News* reported that the PLAAF party committee had directed the "construction of a modernized strategic air force" that is "compatible with China's international position" and capable of "carrying out the historic missions of the armed forces."[7] In this article, author Dong Wenxian—a former researcher at PLAAF headquarters and a member of the PLAAF's Military Theory Studies Expert Group—defined a "strategic air force" as an "integrated air and space force that can play a strategic role in overall national defense, war, and other military struggles, and serves side by side with the army and navy."[8] He explained that this required the PLAAF to have the capability to operate "within the entire strategic space of our country." It defined *strategic space* in four parts: (1) "territory," including air, land, and sea; (2) "seas with resource development and administrative rights," including the exclusive economic zone, continental shelf, and other maritime space; (3) "open land, open space, and outer space"; and (4) "other regions that involve national interests." On the last two areas, the article noted that "the current trend of development" had led many countries to "occupy strategic space outside their own territories."[9]

By late 2008, the PLAAF leadership had directed the development of the ability to carry out a broad range of operations both within and outside the country. In December 2008, the 10th CCP Committee of the PLAAF stated that the objective of the PLAAF's construction should be to become a "strategic air force" with "integrated aerospace and offensive and defensive

---

[6] Hartnett, 2014, p. 55.

[7] Dong Wenxian [董文先], "The Expansion of National Strategic Space Calls for a Strategic Air Force" ["国家战略空间扩展呼唤战略空军"], *Air Force News* 《空军报》, February 2, 2008, p. 2.

[8] Dong Wenxian [董文先], 2008.

[9] Dong Wenxian [董文先], 2008.

capabilities that matched the country's international status and met the requirements for safeguarding national security and developmental interests."[10] It called for developing the capability to "create a posture, control a crisis, and win a war in the country's territorial air *and in areas that are related to the nation's interests* [emphasis added]."[11] It also called for having the ability to "cope with multiple security threats, fulfill diversified tasks, and acting as a pillar for the country's military power."[12] In addition to the focus on offensive and defensive operations in the "strategic air force" concept, according to the most recent version of SMS published in 2013, one of the PLAAF's five "strategic missions" includes the ability to conduct emergency and disaster relief operations. SMS 2013 also called for the PLAAF to strengthen its "strategic transport forces," including its "air strategic delivery capability" and airborne forces.[13]

Military thinkers in the mid-2000s began to trace the implications for the development of long-distance deployment. One of the key concepts concerned "strategic [power] delivery" ["战略投送"], which consists of strategic actions by ground, maritime, and air forces to transport military forces across large distances to deliver military power for the purposes of accomplishing a strategic objective. An article in *PLA Daily* noted that the "obvious insufficiency" in strategic delivery had become a "short plank" that impeded the military's development.[14] It outlined the need to develop the capability to "move military forces far away and fast" to overcome the problem of "being unable to respond effectively to use military forces."[15] As a means of implementation, the article recommended that the military "give priority to elite units" to "perfect support mechanisms" and explore methods of generating strategic lift. It also recommended combining civilian with military resources to meet needs.

To develop and hone force projection capabilities, the *PLA Daily* article recommended the military "go abroad" for exercises to train with foreign military forces, "especially with those in developed countries."[16] Again reflecting the domestic and international application of the "strategic delivery" concept, the article also recommended carrying out long distance "strategic delivery exercises" ["战略投送演练"] within China.[17]

---

[10] Wang Ran [王冉] and Tian Wei [田炜], "Tenth Air Force CCP Committee Holds Its 11th Plenary (Enlarged) Meeting in Beijing" ["空军党委十届十一次全体（扩大）会议在京召开"] *Air Force News* 《空军报》, December 30, 2008, p. 1.

[11] Wang Ran [王冉] and Tian Wei [田炜], 2008.

[12] Wang Ran [王冉] and Tian Wei [田炜], 2008.

[13] See the PLA Academy of Military Science Military Strategy Research Department, ed., *The Science of Military Strategy* 《战略学》, Beijing: Military Science Press, 2013, pp. 222–224.

[14] Zhao Zongqi [赵宗岐], "Put the Construction of Strategic Delivery Capability in an Important Position" [把战略投送能力建设摆到重要位置], *PLA Daily* 《解放军报》, April 23, 2009, p. 10.

[15] Zhao Zongqi [赵宗岐], 2009.

[16] Zhao Zongqi [赵宗岐], 2009.

[17] Zhao Zongqi [赵宗岐], 2009.

Context: The U.S. Experience as an Expeditionary Air Force

China's increasing interest in expeditionary capabilities reflects the country's maturing as a great power and the globalization of its economic and strategic interests. To protect its overseas interests, China aims to promote social stability in countries featuring large Chinese investments, counter threats to its overseas personnel and investments, provide HA/DR to facilitate the restoration of regional stability, and if necessary, evacuate overseas citizens who face a crisis.

However, China's initial development of a limited expeditionary capability in the 2000s should be viewed in historical perspective. The global leader in airpower, the United States, began operating aviation forces in an expeditionary mode in the early 1900s—nearly a century before the Chinese. By the close of World War I, the U.S. Army Air Service had stationed 740 airplanes in Europe. During World War II, the U.S. Army Air Force expanded to include 80,000 airplanes, many of which operated in an extensive array of foreign bases in the European and Asian theaters.[18]

Despite the drawdown in forces following the end of World War II, the United States has maintained a global network of overseas air bases. The U.S. Air Force (USAF) developed extensive experience with flying, transporting, and fighting on a large and continuous scale, over many years, and in many different types of environments. U.S. forces operated bombers, fighters, transport aircraft, and many other types of aircraft in the Korean War, the Vietnam War, and many other combat and nonwar missions. Following the end of the Cold War, U.S. expeditionary operations have been characterized by joint deployments that have harnessed the use of airpower and featured the use of precision-guided weapons. The USAF also continues to routinely conduct operations simultaneously in different regions of the world.

Some of the most impressive advantages enjoyed by the USAF center on the infrastructure available for the operation of aircraft around the world on a sustained basis. The United States has developed an elaborate network of military aviation bases around the world, most of which are located within countries that have a military alliance with the United States. This network of bases enables the U.S. military to rapidly deploy a broad variety of aircraft abroad. Moreover, the United States has prepositioned stocks and developed the ability to transport large amounts of fuel, munitions, and supplies around the world. This robust global logistics system enables the USAF to carry out operations at a sustained, high level. Years of experience with foreign cultures have also helped the USAF operate comfortably with a wide array of partners around the world.

In research on U.S. air operations, Chinese military analysts have discussed the ability of the U.S. armed forces to rapidly deploy to a theater of conflict and scale up personnel, weapon, and equipment to carry out sustained operations.[19] These analysts note that China lacks the

---

[18] U.S. Air Force Historical Studies Office, "Evolution of the Department of the Air Force," May 4, 2011; Stacie L. Pettyjohn, *U.S. Global Defense Posture: 1783–2011*, Santa Monica, Calif.: RAND Corporation, MG-1244-AF, 2012.
[19] Military Training Department of the General Staff of the Chinese People's Liberation Army, *Research into the*

infrastructure, basing access, and experience to operate on a sustained basis in foreign countries. PLAAF forces that depart the country for any mission must carry out extensive preparation and bring much more equipment than is typically the case for U.S. forces. China's approach to expeditionary activity reflects these realities. The PLAAF's international activity tends to be of a far smaller scale, limited duration, and limited operational capability compared to that of U.S. forces.

Kosovo War, Beijing: Liberation Army Publishing House [解放军出版社], 2000; Wang Yongming, Liu Xiaoli, and Xiao Yunhua, *Research into the Iraq War*, Beijing: Liberation Army Publishing House [解放军出版社], March 2003.

2. PLAAF Expeditionary Aircraft and Units

This chapter surveys the units and aircraft that have been most active in expeditionary activities. While a growing variety of units and aircraft have ventured abroad for exercises and other activities, the most active units and in-demand aircraft remain the strategic transport units and their *Ilyushin*-76 (Il-76) heavy transport aircraft.[20] While principally focused on aviation units, this chapter will also review the expeditionary activities of the PLAAF's 15th Airborne Corps. The information cut-off date for overseas activities discussed in this and subsequent chapters is October 2016.

PLAAF fixed-wing aircraft, including fighters, fighter-bombers, and transport aircraft, have deployed long distances within China's borders as well as internationally.[21] Although various combat aircraft have traveled internationally to conduct exercises with foreign militaries, PLAAF transport aircraft have featured especially prominently in missions abroad. Table 2.1 shows the quantity, range, and transport capacity of PLAAF transport aircraft known to have deployed abroad as well as newer medium- and heavy-transport aircraft systems, including the Y-20 large military transport aircraft, which entered service in July 2016.[22]

The PLAAF's seven former Military Region Air Forces (MRAFs) were consolidated to five Theater Command Air Forces (TCAFs) in early 2016. The PLAAF's primary transport units are the 13th Transport Division from the former Guangzhou Military Region and the 4th Transport Division from the former Chengdu Military Region. A separate former Beijing MRAF division is responsible for transportation of Very Important Personnel (VIPs).[23] Not included in this chart are most of the PLAAF's light aircraft, including Y-5s, Y-7s, Y-11s, and Y-12s, many of which are assigned to independent transportation units in the different MRAFs (now TCAFs). None of

---

[20] China made arrangements to acquire ten Il-76s from Russia and four from Uzbekistan in the 1990s (the latter to use for an airborne early-warning aircraft). In 2005, China placed an order from Russia for at least 30 new Il-76TD aircraft in 2005, but production suffered delays and the order was likely not filled, even in part. After these setbacks, China purchased additional refurbished Il-76s. See "China Ahead in Asian States' Post–Cold War Battle," *Jane's Defence Weekly*, September 25, 1993; "In Brief Chinese AEW-A Correction, *Jane's Defence Weekly*, September 7, 1997; "Russia and China Agree Sale of Transport Aircraft and Fuel Dispensers," *Jane's Defence Industry*, September 9, 2005; Jon Grevatt, "China's Il-76/Il-78 Order from Russia Faces Setback," *Jane's Defence Industry*, December 18, 2008; "Il-76MD Transport Plane Delivered to China," *The Moscow Times*, January 24, 2013.

[21] PLAAF helicopters have also deployed abroad, but their activities are not covered in this analysis. A PLAN Y-9 transport aircraft apparently traveled to Thailand in September 2016 to support the Association of Southeast Asian Nations (ASEAN) Defense Ministers' Meeting Plus AM-HEx 2016 exercise. See "Chinese Troops Arrive in Thailand to Participate in the Second ASEAN Defense Ministers' Meeting Plus Humanitarian Assistance Disaster Relief and Military Medicine Joint Drill" ["中国军队抵达泰国参加第二次东盟防长扩大会人道主义援助救灾与军事医学联合演练"], Ministry of National Defense, September 1, 2016.

[22] "Chinese Large Freighter Plane Enters Military Service," Xinhua [新华] in English, July 7, 2016.

[23] This was the 34th VIP Transport Division. There were also transport regiments in the special mission divisions located in the former Shenyang, Nanjing, and Chengdu MRAFs.

these have been observed to deploy abroad, with the possible exception of the two Y-7s and six Y-8s that are organic to the 15th Airborne Corps. A comparison with U.S. forces can provide a sense of the limited Chinese inventory. The USAF alone has about 265 heavy- and 428 medium-lift aircraft.[24]

Table 2.1. Select PLAAF Medium- and Heavy-Transport Aircraft

Aircraft Name	Quantity	Range with Payload	Transport Capacity (Max Payload)	Total Quantity
Y-20	3	3,700–5,200 km (estimated)	Heavy (66 metric tons)	
Il-76MD/TD	20	3,648–4,000 km	Heavy (47–50 metric tons)	
Y-9	12+	2,200 km (estimated)	Medium (20 metric tons)	
Y-8	30	1,272–2,083+ km	Medium (20 metric tons)	
				65+

SOURCES: Quantities are from International Institute for Strategic Studies, *The Military Balance 2017*, London: Routledge, 2017. Range and capacity data are from IHS Jane's database; these data may include multiple aircraft variants for each model and do not necessarily represent the specific variant that has deployed abroad. Table does not include 15th Airborne Corps, personnel, or VIP transport aircraft. Other services have limited numbers of medium- and light-transport aircraft; both the PLAA and the PLAN have a small inventory of Y-7s and Y-8s, and the PLAA reportedly received its first Y-9 in December 2016. See Sun Xingwei [孙兴维] and Li Yunpeng [李云鹏], "First Y-9 Type Transport Aircraft Officially Enters Services with Army Aviation Forces" ["首架运-9 型运输机正式列装陆军航空兵部队"], *China Army Online* 《中国陆军网》, December 23, 2016.

The Y-20 will further bolster the PLAAF's transport capacity and support PLA-wide contingency operations. Media reports indicate that the first Y-20s have been assigned to units in the former Chengdu MRAF.[25]

Profile: The 13th Transport Division

The 13th Transport Division [运输机师], stationed in the former Guangzhou MRAF, has played a leading role in the PLAAF's expeditionary activities. The PLAAF established the division in 1951.[26] It has three regiments: The 37th Air Regiment and the 38th Air Regiment are equipped with medium-capacity transports, and the 39th Air Regiment is equipped with the current mainstay of the PLAAF's heavy-lift aircraft, the Il-76.[27] The 13th Transport Division

[24] Of these, approximately 239 heavy- and 145 medium-lift aircraft are in the active force. C-5 data are a projection for fiscal year 2017 from January 2014; C-17 data are from October 2015; and C-130 data are from May 2014. See "C-5 A/B/C Galaxy and C-5M Super Galaxy," January 2014; "C-17 Globemaster III," October 2015; "C-130 Hercules," May 2014.

[25] Steven Jiang, "China's Military Gets Boost with Huge New Transport Plane," *CNN*, July 7, 2016.

[26] Kenneth W. Allen, Glenn Krumel, and Jonathan D. Pollack, *China's Air Force Enters the 21st Century*, Santa Monica, Calif.: RAND Corporation, MR-580-AF, 1995, p. 40.

[27] Andreas Rupprecht and Tom Cooper, *Modern Chinese Warplanes: Combat Aircraft and Units of the Chinese Air Force and Naval Aviation*, Houston: Harpia Publishing, 2012, p. 189. For more information on some of the female

carries out diverse missions including HA/DR, paradrop, and cargo delivery. The division also has a small number of all-female air crews.[28]

The 13th Transport Division is regularly described as the country's premier unit for overseas air transport, personnel recovery, participation in international exercises, and delivery of HA/DR supplies. The unit has delivered relief materials to a variety of countries on China's periphery, including Afghanistan, Pakistan, Mongolia, and Myanmar.[29] Media profiles of the 13th Transport Division describe an extensive training program. One report stated that the division's pilots frequently train in a diverse array of conditions and can log thousands of kilometers annually in training and in real-world missions. In April 2011, one of its regiments reportedly carried out its first cargo airdrop after flying over water at night and under tactical conditions.[30]

Chinese military media reporting has also provided some insight into how pilots of the 13th Transport Division operate. A 2012 article in *Air Force News* profiled one regiment from the division, most likely the 39th Regiment. The regiment reportedly has two crews per aircraft to ensure readiness for rapid deployment. The report noted that the regiment had adopted "new modes of flight organization and command."[31] It explained that personnel had moved away from "babysitter-style guidance" toward a mode in which pilots conducted "autonomous preparations, takeoff and landing, coping with risks, and task planning."[32] The report stated that this involved "emergency deployment support modes" in which the crews "directly make preparations onboard for instant deployment, ad-hoc on the spot course correction, and timely arrival at the destination."[33] The goal of the reorganized command mode is to enable the aircrews to deploy and take off as soon as possible after receiving orders.[34]

This level of pilot autonomy reflects, in part, a broader trend in the PLAAF in which pilots have been expanding the levels of free air combat and pilot autonomy. However, the focus on developing pilot autonomy may also reflect the more solitary, expeditionary nature of many of the missions undertaken by the transport division. Media reports note that the regiment levied requirements to increase a more-autonomous mode of operations in the early 2000s, well before

---

aircrews in the division, see Kenneth Allen and Emma Kelly, "China's Air Force Female Aviators: Sixty Years of Excellence (1952–2012)," *China Brief*, Vol. 12, No. 12, June 22, 2012.

[28] Allen and Kelly, 2012.

[29] Yang Yongsheng [杨永生], Zhu Zhanghu [朱章虎], and Zhao Lingyu [赵凌芋], "Training that Builds 'Iron Wings'" ["练就'铁翅膀'"], *Air Force News* 《空军报》, August 21, 2012, p. 1.

[30] Dong Ruifeng [董瑞丰] and Guo Hongbo [郭洪波], "Sharp Weapons of a Big Power: Visit to an Air Force Strategic Transport Aircraft Unit" ["大国利器: 探访空军战略运输机部队"], *Outlook* 《了望 》, No. 34, August 20, 2012, pp. 48–49.

[31] Yang Yongsheng [杨永生], Zhu Zhanghu [朱章虎], and Zhao Lingyu [赵凌芋], 2012.

[32] Yang Yongsheng [杨永生], Zhu Zhanghu [朱章虎], and Zhao Lingyu [赵凌芋], 2012.

[33] Yang Yongsheng [杨永生], Zhu Zhanghu [朱章虎], and Zhao Lingyu [赵凌芋], 2012.

[34] Yang Yongsheng [杨永生], Zhu Zhanghu [朱章虎], and Zhao Lingyu [赵凌芋], 2012.

the PLAAF highlighted greater autonomy as a requirement for its combat pilots.[35] One 2005 article described how one of the 13th Transport Division's regiments operated in a manner that de-emphasized the role of ground control. It explained that a group of six large transports trained according to the principle of "three withouts," defined as "1) without command guidance from the ground; 2) without any weather or wind data for the drop zone; 3) without any manual signals or markers for the drop zone."[36] To do this, the regiment prepared training topics, conducted computer-simulated training events, and gathered and prepared relevant data to plan the mission.[37]

Profile: PLAAF Airborne Troops

The PLAAF's 15th Airborne Corps is China's primary strategic airborne unit. The corps includes the 43rd, 44th, and 45th Divisions. It is headquartered in Xiaogan, Hubei.[38] The 15th Airborne Corps has its own organic transport units, consisting of Y-7s and Y-8s. However, Il-76s from the 13th Transport Division also support airborne operations. Because of the limited availability of heavy transport, only one division is likely capable of deploying rapidly across the country at a time.

While principally a rapid reaction unit for missions on the country's periphery, airborne troops and transport planes from the 13th Transport Division have begun to operate together in exercises abroad. The increasing international activity builds on years of efforts to extend the operational range of airborne troops through paradrop operations and delivery of cargo by air. For example, the 13th Transport Division has practiced air-dropping cargo and paratroopers since at least the mid-2000s. Airborne troops have increased training in long-distance exercises as well. In a September 2010 exercise, paratroopers and their assault vehicles parachuted from airplanes to land on a plateau in China's northwest.[39] In recent years, the Airborne Corps has also been practicing parachuting at altitudes of over 4,600 meters. This domestic training has reportedly increased the ability of airborne forces to operate in a broad variety of climates and terrain. According to one report, "all the servicemen of the Airborne Corps, irrespective of gender and age, from generals to the rank and file, can jump from five aircraft models over nine types of terrain and open double parachutes, among other things, and are ready to perform

---

[35] Andrew Erickson, "China's Modernization of Its Air and Naval Capabilities," in Ashley Tellis and Travis Tanner, eds., *Strategic Asia 2012–13: China's Military Challenge*, Washington, D.C.: National Bureau of Asian Research, 2012, p. 76.

[36] Fan Haisong [范海松] and Ding Yanli [丁艳丽], "Capability of a Guangzhou MR Transport Aviation Division's Long Range Airlift and Long Range Mobility Operations Ascends" ["广州运输航空兵某师远程空运和远程机动作战能力跃升"], *Air Force News* 《空军报》, March 26, 2005, p. 1.

[37] Fan Haisong [范海松] and Ding Yanli [丁艳丽], 2005.

[38] Dennis J. Blasko, *The Chinese Army Today: Tradition and Transformation for the 21st Century*, 2nd ed., New York: Routledge, 2012, p. 104.

[39] "Chinese Il-76 to Conduct Heavy-Equipment Air Dropping," *China Military Online* in English, August 22, 2014.

various missions at any time."[40] Its overseas activities have to date only consisted of paradrops of troops and equipment in bilateral and multilateral exercises (described later in this report).

Profile: Bayi Aerobatics Team

The PLAAF's demonstration team, the Bayi Aerobatics Team [八一飞行表演队], is a unit consisting mainly of fighter aircraft that has likely deployed abroad more often than other fighter aircraft units, albeit exclusively in a noncombat, military diplomacy role. The Bayi team was founded in 1962 and currently flies J-10 fighter aircraft. It has performed for a wide variety of audiences within China, including for foreign delegations and at National Day military parades.[41] In 2013, it expanded to events outside of China by performing at the Moscow Air Show, followed by performances in Malaysia and Thailand in 2015. While the Bayi team primarily serves in a military diplomacy role, traveling Bayi units must undertake similar responsibilities to other units that deploy overseas, such as planning and coordinating international flight paths and carrying out operations, logistics, and maintenance at foreign locations.

---

[40] "Heeding Party Commands, Serving the People, and Fighting Bravely and Skillfully" ["听党指挥 服务人民 英雄善战"] series, CCTV-7 "*Military Report*" [军事报道] program, December 7, 2006.

[41] For more information on the Bayi team, see Michael S. Chase, Kenneth W. Allen, and Benjamin S. Purser III, *Overview of People's Liberation Army Air Force "Elite Pilots,"* Santa Monica, Calif.: RAND Corporation, RR-1416-AF, 2016.

3. Key Trends and Themes in PLAAF Deployments

Overall, PLAAF deployments abroad have primarily consisted of two types of operations: participation in exercises with foreign militaries and HA/DR operations. The PLAAF has also carried out an NEO and begun to participate in international competitions and air shows. However, it has not played a role in United Nations peacekeeping operations so far. Although the sample size is small, the PLAAF's rate of conducting international deployments appears to be on the rise, particularly for HA/DR operations.

As most of these operations have involved the transportation of supplies and occasionally personnel from China, the 13th Transport Division has played an especially prominent role in the PLAAF's overseas operations. However, future deployments may see greater diversity in aircraft as the PLAAF grows more confident in operating abroad.

Major Types of Deployments

As the requirements for the PLAAF to conduct expeditionary operations have grown, so too have the tasks it has been called on to perform. This section covers the range of operations carried out by the PLAAF in recent years. Since the Chinese military has not engaged in major combat since its conflict with Vietnam in 1979, all of the PLAAF's overseas deployments have been for missions other than war.

Bilateral and Multilateral Exercises

Since the Peace Mission 2007 exercise, the PLAAF has participated in exercises outside of China with the militaries of Shanghai Cooperation Organization (SCO) member countries, including forces from Kazakhstan, Kyrgyzstan, Russia, Tajikistan, and Uzbekistan. Peace Mission 2013 also involved PLAAF deployments abroad. In other exercises, the PLAAF has deployed units to Turkey, Pakistan, Malaysia, and Thailand. Through these exercises, the PLAAF has learned to navigate outside the country, work with foreign partners, and carry out limited operations on an expeditionary footing. Table 3.1 summarizes these exercises.

Table 3.1. Known PLAAF Participation in Exercises Outside of China

Year	Exercise (Location of Deployment)	Partners	PLAAF Complement (Unit's MRAF, If Known)
2007	Peace Mission (Russia)	Kazakhstan, Kyrgyzstan, Russia, Tajikistan, Uzbekistan	Eight JH-7s (NJ MRAF 28th Air Division), six Il-76s (GZ MRAF)
2010	Anatolian Eagle (Turkey)	Turkey	Four Su-27s/J-11s, one Il-76 (GZ MRAF)
2011	Shaheen I (Pakistan)	Pakistan	Su-27s
2013	ADMM-Plus HADR/MM Ex (Brunei)[a]	N/A	One Il-76
2013	Peace Mission (Russia)	Russia	Five JH-7As (SY MRAF), two Il-76s (GZ MRAF)
2014	Shaheen III (Pakistan)	Pakistan	J-10s, J-7s
2015	Peace and Friendship (Malaysia)	Malaysia	Four transport aircraft, including at least one Il-76 (GZ MRAF) and one Y-8C
2015	Falcon Strike (Thailand)	Thailand	Six J-11As; 180 Chinese officers and pilots
2016	Shaheen V (Pakistan)	Pakistan	Unknown
2016	AM-HEx (Thailand)[b]	Thailand, Russia	PLAAF General Hospital personnel

SOURCE: Chinese articles cited in this section ("Bilateral and Multilateral Exercises").
NOTE: Does not include air shows/competitions, activities of the 15th Airborne Corps, or PLAA or PLAN assets.
[a] ADMM-Plus HADR/MM Ex = ASEAN Defense Ministers' Meeting (ADMM)-Plus HA/DR and Military Medicine Exercise.
[b] AM-HEx = ADMM-Plus Military Medicine–HA/DR Joint Exercise.

Peace Mission SCO Exercises

Peace Mission 2007

Peace Mission 2007 marked the first time that the PLAAF sent combat aircraft abroad as well as the first SCO exercise involving all six member states at that time (China, Russia, Kazakhstan, Kyrgyzstan, Tajikistan, and Uzbekistan). The exercise was described by Chinese media as a "joint antiterror drill" that ran from August 9 through August 17.[42] The PLA deployed fixed- and rotary-wing aircraft to Shagol Airfield in the Chelyabinsk region of Russia, north of Kazakhstan. The 46 participating Chinese aircraft included eight JH-7 fighter-bombers from the 28th Air Division (of the former Nanjing MRAF) and six PLAAF Il-76s from the 13th Transport Division, as well as 32 PLA Army Aviation helicopters. PLAAF airborne troops from the 15th Airborne Corps also participated, though in a China-hosted portion of the exercise. Reflecting the event's political importance, PLAAF General Xu Qiliang, then one of the deputy chiefs of the PLA General Staff, served as the Chinese commander-in-chief for the exercise.[43]

---

[42] "SCO Leaders Observe Joint Anti-Terror Drill," Xinhua [新华] in English, August 17, 2007.

[43] Le Tian, "Peace Mission Exercises Get Under Way," *China Daily Online* in English, August 7, 2007, p. 3.

The PLAAF aircraft flew 2,700 kilometers to participate.[44] The Chinese aircraft conducted reconnaissance, "firepower attacks," and airdrops. An article by Li Daguang, a professor at the PLA National Defense University, noted that the Il-76 transport aircraft completed 18 sorties in the exercise.[45] The PLAAF's transport aircraft worked with their Russian counterparts during one portion of the exercise, taking off in one-minute intervals to conduct consecutive airdrops.[46] The Chinese aircraft dropped 20 pieces of "technical equipment" as well as 200 paratroopers.[47] One article noted that the transport regiment had shortened the intervals between airdrops and parachute landings by one-third in 2005, implying that these skills were put to the test during the exercise.[48] During the drops, the transport aircraft from both countries carried out three consecutive airdrop and paradrops, despite stormy weather conditions.

Chinese coverage of the exercise article implied that the PLAAF only took the most-reliable and capable pilots and support personnel. One article noted that selected pilots were "special grade" and that "fewer than 40 aircraft maintenance personnel were responsible for work that was originally performed by nearly 200 personnel."[49] It described all of the participating personnel in the exercise as "elite troops" ["精兵强将"].[50]

Peace Mission 2013

For Peace Mission 2013, which was also described as a counterterrorism exercise, five JH-7A fighter-bombers from the former Shenyang MRAF and two Il-76s traveled to Russia.[51] They took off from Siping City in Jilin Province and arrived at Shagol Airfield after three stops. Overall, the aircraft flew 6,098 kilometers, 2,800 kilometers of which were in foreign airspace.[52]

Exercise participants learned that limitations in the types of ordnance used by each side impaired the ability to carry out combined operations against shared targets. One Russian article observed "the strike power of the aviation could have been stronger" during this portion of the exercise, but

> [a]fter consulting with the military aviation specialists of the two countries, it was revealed that the arsenal of the Chinese fighters did not include ground

---

[44] "'Peace Mission 2007' to Test Remote Mobile Ability of Chinese Forces," *PLA Daily* 《解放军报》.

[45] Li Daguang [李大光], "PLA 'First-Time' Achievements in 'Peace Mission 2007' Exercise" ["解放軍在'和平使命—2007'演習中的'第一次'"], *Wen Wei Po* 《文匯網》, August 24, 2007.

[46] Dong Ruifeng [董瑞丰] and Guo Hongbo [郭洪波], 2012.

[47] Li Daguang [李大光], 2007.

[48] Dong Ruifeng [董瑞丰] and Guo Hongbo [郭洪波], 2012.

[49] Li Daguang [李大光], 2007.

[50] Li Daguang [李大光], 2007.

[51] Zhang Haiping, Zhang Feng, Zhang Zhe, Ouyang Dongmei, and Zhang Tao, "Peace Mission 2013: China-Russia Joint Anti-Terrorism Exercise," *China Military Online* in English, August 15, 2013.

[52] "PLAAF Combat Group Holds First Flight Training" ["中方空军战斗群首次飞行训练"], CCTV-13 News [CCTV-13 新闻] program, August 7, 2013.

> support ammunition. Their main on-board combat stores were normal and guided aviation bombs. However, it was not safe to use this kind of ammunition from the 200-meter altitude of the Russian bombers. As a result, corrections had to be made in the originally developed plan for the joint bombing of the Su-25Ms and the JH-7As that somewhat changed the sequence for executing the tasks. For the sake of fulfilling the exercise scenario, the Russian side retained the use of the ground-support munitions at the 200-meter altitude, and the Chinese pilots attacked the targets using regular ammunition from high altitudes.[53]

Another shortcoming involved lack of direct radio communications between Russian and Chinese pilots. The article gave an example in which a PLAAF pilot might execute a turn and report it to the command post, but the Russian pilot would not receive this information until it was received at the Chinese command post, translated, and finally forwarded by the Russian flight leaders.[54] To get around the delay, the two sides worked out gestures to use in the air during the exercise.

Besides the ground forces, the Chinese side included an "air force battle group" (50 people), a "comprehensive support group" (196 people), an "exercise-directing unit" (20 people), and an "operations-command unit" (30 people).[55] An advance group of 33 PLA members coordinated arrangements regarding "foreign affairs, operations, political work, logistics, armaments, and other relevant areas."[56]

Other Recent Experience with the Russian Military

In August 2015, China and Russia held Joint Sea–2015 (II), a combined naval exercise in the Peter the Great Gulf that involved the Russian Navy and the PLAN as well as an "air defense topic" featuring the PLAAF (the first Joint Sea–2015, held in May, only involved the PLAN). The PLAAF aircraft involved flew out of Chinese airfields and do not appear to have operated out of Russian facilities for the exercise.[57] However, the exercise is still noteworthy as it was reported to be the first international exercise involving both PLAN and PLAAF participants.[58] Two J-10s, two JH-7s, and one KJ-200 formed a "Chinese air combat formation," in which the KJ-200 provided airborne early warning for the fighters. During the "first"-ever combined

---

[53] Yuriy Belousov [Юрий БЕЛОУСОВ], "In Merit There is Honor" [По заслугам и честь], *Krasnaya Zvezda Online* 《Красная звезда》, August 22, 2013.

[54] Yuriy Belousov [Юрий БЕЛОУСОВ], 2013.

[55] Zhang Haiping et al., 2013.

[56] Zhang Haiping et al., 2013.

[57] Xu Xieqing [徐叶青] and Ren Xu [任旭], "Sino-Russian 'Joint Sea-2015 (II)' Live Troop Exercise Fires First Shots" ["中俄 '海上联合-2015（Ⅱ）' 实兵演习打响"], *China Military Online* 《中国军网》, August 24, 2015; "Chinese Air Force Dispatches 3 Types, 5 Combat Aircraft to Participate in Sino-Russian Joint Military Exercise" ["中国空军出动 3 型 5 架战机参加中俄联合军演"], Central Government Web Portal [中央政府门户网站], August 24, 2015.

[58] Zhang Zhongkai [张钟凯] and Wu Dengfeng [吴登峰], "Military Experts Explain Highlights of Sino-Russian 'Joint Sea-2015 (II)' Exercise in Detail" ["军事专家详解中俄 '海上联合-2015 (II)' 演习亮点"], Xinhua [新华网], August 25, 2015.

Chinese and Russian "naval three-dimensional landing" exercise, the two types of PLAAF fighters took off separately, "executed firepower support across the water of the landing location," and conducted simulated attacks "at the landing highland."[59] Paratroopers also participated, although it is unclear if these units were from Russia, China, or both countries.

Anatolian Eagle

Four Su-27s and at least one Il-76 traveled from Xinjiang to Turkey to participate in Anatolian Eagle in 2010.[60] The exercise marked the first time that the PLAAF conducted air maneuvers with a member of the North Atlantic Treaty Organization, although Turkey flew F-4s rather than F-16s during the exercise, possibly in response to U.S. concerns.[61] To travel the 3,500 kilometers to Turkey, the aircraft reportedly refueled in Pakistan and Iran en route and in Iran on the return route.[62] Coordinating these refueling stops likely posed a logistical and diplomatic test prior to the exercise.

Chinese media coverage of the exercise was limited, but Turkish media reported that the participants conducted a dogfight over Konya in central Anatolia.[63] Shortly after the exercise was completed, Prime Minister Wen Jiabao traveled to Ankara, marking the first visit by a Chinese premier to Turkey in eight years. During his trip, the countries announced a "strategic partnership" between China and Turkey as well as a goal of expanding bilateral trade to $50 billion by 2015 and $100 billion by 2020.[64]

---

[59] Hu Shanmin [胡善敏], Xu Miaobo, Li Wei [李伟], Xia Chenghao [夏程浩], and Cheng Pengcheng, "China and Russia Conclude 'Joint Sea 2015 (II)' Exercise," CCTV News "Focus On" program, August 28, 2015; Hu Shanmin [胡善敏], Kou Lihua [寇利华], Li Wei [李伟], Xia Chenghao [夏程浩], and Li Xiulin [李秀林], "Four Subjects a Day Set Record of Highest Density of Drills" ["一天联演四课目 演习密度最高"], CCTV News "Focus On" program, August 28, 2015; "Air Force Aviation Force Flies Out into the World for Training, What News Has Been Transmitted?" ["空军航空兵飞出国门联训，传递了什么信息？"], *China Military Online*《中国军网》, April 9, 2016.

[60] "PLAAF Su-27 Fighters Refueled in Iran While Traveling to Turkey for Exercise" ["中国空军苏-27 战机赴土耳其途中曾在伊朗加油"], *Dongfang Online* 《东方网》, October 11, 2012.

[61] Jim Wolf, "China Mounts Exercise with Turkey, U.S. Says," Reuters, October 8, 2010.

[62] "PLAAF Su-27 Fighters Refueled in Iran While Traveling to Turkey for Exercise" [中国空军苏-27 战机赴土耳其途中曾在伊朗加油], 2012.

[63] "In Erdogan's Skies" ["Nei cieli di Erdogan"], *Milan Il Foglio* [*The Milan Paper*], October 14, 2010, p. 1; "Turkey, China Conduct Joint Air Maneuvers," *Istanbul Today's Zaman Online* in English, September 30, 2010. After rumors swirled online that Turkish air forces outperformed Chinese forces during the exercise, the PLAAF issued a statement debunking the claims. Yang Tiehu [杨铁虎] "Air Force Confirms: 'PLAAF Defeated by Turkey in Air Combat Exercise' Is Simply a Rumor" ["空军证实: '中国空军空战惨败土军' 纯属谣言"] *People's Daily* 《人民日报》, October 15, 2010.

[64] For more information, please see Chris Zambelis, "Sino-Turkish Strategic Partnership: Implications of Anatolian Eagle 2010," *China Brief*, Vol. 11, No. 1, January 14, 2011.

Shaheen Series with Pakistan

The Shaheen series of exercises with Pakistan has thus far included three exercises held in Pakistan in 2011, 2014, and 2016 and two in China in 2013 and 2015. Chinese media reported a rare copiloting of a Chinese Su-27 by both Chinese and Pakistani pilots during Shaheen I (2011). One article noted that, for the PLAAF, training exercises with the Pakistani Air Force helped China "understand the tactical style and equipment performance of imaginary enemies that may engage in warfare with China in the future."[65] For Shaheen III (2014), China reportedly sent J-10 and J-7 fighter aircraft to Pakistan. Chinese media described the exercise as a "multidimensional joint military drill, which enhanced cooperation and demonstrated skills through air confrontation in a highly simulated war environment."[66] Shaheen V ran from April 9 to April 30, 2016, and reportedly included training events involving combat pilots, air traffic controllers, and ground crews.[67]

Southeast Asia Exercises

During the ASEAN Defense Ministers' Meeting (ADMM)-Plus HA/DR and Military Medicine Exercise in June 2013, Bruneian media reported that a Chinese Il-76 carried a 15-member crew and 34 personnel to Brunei for the exercise, along with HA/DR medical equipment.[68] It is unclear which PLA service the personnel came from.

China's first bilateral military exercise with Malaysia, "Peace and Friendship 2015," was largely focused on naval training topics. China sent three ships, four transport aircraft (including at least one Il-76 and one Y-8C), three helicopters, and 1,160 troops from the PLAA, PLAN, and PLAAF to participate.[69] One Il-76 and one Y-8C participated in a drill to locate and rescue a kidnapped vessel.[70] Chinese media coverage of the closing ceremony noted that the exercise was China's largest with an ASEAN country to date, and Yi Xiaoguang, deputy chief of the General Staff Department, stated that "China is committed to promoting cooperation and exchanges with

---

[65] "Air Forces of China and Pakistan Benefit from Joint Training," *China Military Online* in English, September 17, 2015.

[66] "Air Forces of China and Pakistan Benefit from Joint Training," 2015.

[67] "China-Pakistan Air Forces Hold Shaheen V Joint Exercise in Pakistan" ["中巴空军在巴基斯坦举行 '雄鹰-V' 联合训练"], Ministry of National Defense, April 9, 2016; "Pak-China Air Exercise Shaheen V Begins," *Radio Pakistan* in English, April 15, 2016; "Air Force Aviation Force Flies Out into the World for Training, What News Has Been Transmitted?" [空军航空兵飞出国门联训，传递了什么信息？], 2016.

[68] Bandar Seri Begawan, "More Troops Arrive for Military Exercises," *Brunei Times Online* in English, June 10, 2013.

[69] "China-Malaysia Hold Peace and Friendship 2015 Joint Exercise" ["中马将举行 '和平友谊-2015' 联合军事演习"], Ministry of National Defense, August 27, 2015; Dzirhan Mahadzir, "China, Malaysia Hold First Bilateral Field Drills," *Jane's Defence Weekly*, September 18, 2015.

[70] Qu Yantao [曲延涛], Yang Qinggang [杨清刚], and Xu Xiaolong [徐小龙], "Peace and Friendship 2015 China Malaysia Maritime Drill," *China Military Online* 《中国军网》, September 20, 2015.

defense departments and armed forces from all ASEAN countries . . . to jointly face the new challenges and threats and maintain regional peace."[71]

In November 2015, the PLAAF participated in its first bilateral air force exercise with Thailand at Korat Airbase.[72] One hundred and eighty Chinese officers and pilots participated, with six J-11A/Su-27s operating out of a Thai base used by U.S. forces to launch bombing missions during the Vietnam War.[73] Chinese commentators also highlighted the significance of China expanding military exchanges with ASEAN countries as well as a country that has historically had close security ties to the United States.[74]

In September 2016, China sent ground and naval forces as well as a 40-person medical detachment primarily from the PLAAF General Hospital to Thailand for the second ADMM-Plus Military Medicine–HA/DR Joint Exercise.[75] According to one article, the exercise marked the first time PLAAF medical service personnel had participated in an international event. The detachment treated several dozen injured people and performed a dozen surgical operations, some together with Thai and Russian counterparts.[76] They also were the only force to set up a second-class field hospital in addition to host country Thailand.

Airborne Troops' Participation and Cargo Delivery in Exercises

Airborne troops have participated in bilateral and multilateral exercises abroad since 2011. Table 3.2 lists known international exercises undertaken by airborne troops.

---

[71] "China Malaysia Conclude First Joint Military Exercise," Xinhua [新华] in English, September 23, 2015.

[72] "China-Thai Joint Air Force Drill: A Signal for More Open Chinese Air Force," *China Military Online* in English, November 17, 2015.

[73] "China-Thailand 'Falcon Strike 2015' J-11s Encounter 'Gripen'" ["中泰 '鹰击－2015' 歼－11 遇 '鹰狮'"], *CCTV News Online* 《央视网》, November 12, 2015; "PLA Air Force J-11s Visit Thailand to Challenge Gripen Fighters-Reports from the Scene" ["解放军多架歼 11 赴泰叫阵鹰狮战机 现场曝光"], *Global Times* 《环球时报》, November 19, 2015.

[74] "China-Thailand 'Falcon Strike 2015' J-11s Encounter 'Gripen'" ["中泰 '鹰击－2015' 歼－11 遇 '鹰狮'"], 2015.

[75] A PLAN Y-9 transport aircraft apparently traveled to Thailand to support the exercise, but no PLAAF aircraft were mentioned. See "Chinese Troops Arrive in Thailand to Participate in the Second ASEAN Defense Ministers' Meeting Plus Humanitarian Assistance Disaster Relief and Military Medicine Joint Drill" ["中国军队抵达泰国参加第二次东盟防长扩大会人道主义援助救灾与军事医学联合演练"], 2016.

[76] "Second AM-HEx 2016 Joint Exercise Underway in Thailand," *China Military Online* in English, September 8, 2016.

Table 3.2. Known Exercises Airborne Troops Have Participated in Outside of China

Year	Exercise	Location
2011	Divine Eagle/Condor	Belarus
2011	Cooperation	Venezuela
2013	Sharp Knife Airborne	Indonesia
2014	Sharp Knife Airborne	Indonesia
2015	Divine Eagle/Condor	Belarus
2016	International Army Games "Airborne Platoon"	Russia

Peace Mission Series

Airborne troops have participated in the Peace Mission series of exercises, although none of their Peace Mission jumps have yet occurred outside of China's borders. Peace Missions 2005, 2007, and 2014 featured PLAAF airborne jumps in China-hosted events that are summarized below.

PLAAF airborne troops jumped in their first international exercise in the SCO's first Peace Mission exercise from August 18 to August 25, 2005. In the China portion of that event, Il-76s dropped six Chinese ZSL-2000 airborne combat vehicles and 86 personnel, mainly the vehicle crews.[77] The 2007 exercise, which took place between August 9 and August 17, saw the PLAAF airborne troops jump with their Russian counterparts, again in the China portion of the exercise. According to Russian authorities, a total of 240 Russian and Chinese paratroopers and 24 pieces of hardware were airdropped from Russian and Chinese Il-76 aircraft.[78] For the Peace Mission 2014 exercise, the paradrop that again occurred in China focused on greater precision in airdropping troops and vehicles to the drop zone.[79]

Divine Eagle

PLAAF airborne troops trained in Belarus in July 2011 as part of a ten-day counterterrorism drill called "Divine Eagle" (also sometimes translated as "Condor," "Celestial Eagle," or "Dashing Eagle"). Chinese media reported that 83 troops traveled from Urumqi to Belarus aboard at least one Il-76. The Belarusian and PLAAF airborne troops combined parachuting and tactical air-landing operations to encircle and eliminate "terrorist" foes. Chinese media stated that this was the first major combined training exercise for Chinese paratroopers in a foreign

---

[77] For more information, see Martin Andrew, "Power Politics: China, Russia, and Peace Mission 2005," *China Brief*, Vol. 5, No. 20, September 27, 2005.

[78] For more information, see Roger McDermott, "The Rising Dragon: Peace Mission 2007," Jamestown Foundation occasional paper series, October 2007.

[79] "Chinese Il-76 to Conduct Heavy-Equipment Air Dropping," 2014.

country as well as the first exercise between the two countries.[80] In June 2015, PLAAF airborne troops returned to Belarus for a combined counterterrorism exercise.[81]

Cooperation 2011

In fall 2011, 21 paratroopers traveled to Venezuela for the first counterterrorism exercise between the two countries.[82] One *PLA Daily* article stated that the Chinese troops participated in a monthlong drill featuring 17 hours of intensive training a day. Some of the training topics included 400-meter altitude armed parachuting during the day and night and 5,000-meter-altitude "parafoil penetration" behind enemy lines, although it is unclear from what type of aircraft the paratroopers jumped.[83]

Sharp Knife Series with Indonesia

During the one-weeklong drill for Sharp Knife Airborne 2013, Chinese paratroopers trained with Indonesian counterparts for the first time. The exercise was held in Bandung, the capital of West Java Province. The two sides conducted live-fire exercises including air landing, hand-to-hand combat, shooting, obstacle crossing, and antiterrorist search-and-rescue operations. They jumped from at least one Indonesian aircraft.[84] For Sharp Knife Airborne 2014, Chinese troops also reportedly carried out "high-altitude" paradrops.[85]

Armored Games Series with Russia

In July 2016, the PLAAF sent a platoon of airborne troops as one of 17 Chinese military teams to take part in the Russia-hosted International Army Games. The airborne troops competed in events involving parachute drops, driving combat vehicles through obstacle courses, operation of weapons, individual fitness and cross-country races, and the integration of troops, weapons, and tactics. A Chinese media report stated that the troops operated principally with a Mi-8 helicopter.[86]

---

[80] "China-Belarus First Airborne Joint Exercise Concludes Well" ["中国与白俄罗斯首次空降兵联合训练圆满结束"], Xinhua [新华], July 19, 2011.

[81] "Chinese Air Force Visits Belarus to Participate in Joint Counter-Terrorism Exercise (Pictures)" ["中国空降兵赴白俄罗斯参加联合反恐训练(图)"], *China News Online* 《中国新闻网》, June 15, 2015.

[82] News clip of Major General Zheng Yuanlin beginning with "The Joint Drill Between China and Venezuela Is a New Trial Exercise" ["中委两军组织联合训练，这是我们一种新的尝试"], CCTV News Content, November 9, 2011.

[83] "PLA Airborne Commando Returns from Anti-Terrorism Drill," *PLA Daily Online* in English, November 23, 2011.

[84] "China, Indonesia Complete Anti-Terror Exercise," *China Daily Online* in English, November 12, 2013.

[85] "'Sharp Knife Airborne-2014' Exercise Concludes," *China Military Online* in English, November 5, 2014.

[86] "Chinese Paratroopers Prepare for Int'l Army Games 2016," *China Military Online* in English, July 27, 2016.

Humanitarian Assistance/Disaster Relief Operations

PLAAF media describe the response to the May 12, 2008, earthquake in Wenchuan, Sichuan Province, as a key development in the PLAAF's expeditionary activity. One report stated that 88 minutes after the quake, the first Il-76 rushed to Nanyuan Airport in Beijing, and it worked through the night to bring disaster-relief teams and supplies to the affected areas. This Il-76 also conducted multiple air drops from an altitude of 600 meters, "completely without ground-based guidance," as well as a 4,999-meter drop of 15th Airborne Corps paratroopers into the disaster area to establish a communications hotline to help coordinate the arrival of additional personnel.[87] On May 14, 2008, 12 large transport planes departed from three different military airfields to transport 1,500 personnel and more than 100 tons of materiel to the disaster area.

According to Chinese media, the response to the Wenchuan earthquake set several records PLA-wide. One article noted that response was "the largest air transportation operation in the history of PLA disaster rescue and relief."[88] The PLAAF and PLA Army Aviation deployed more than 200 airplanes and helicopters to transport 39,000 people and over 7,700 tons of materiel via more than 5,400 flights.[89] For the PLAAF, the mission set the record for the most planes fielded in one day and for the longest operational deployment in the history of the PLAAF.[90] For the duration of the relief operation, 60 transport planes and helicopters ferried 15,000 relief personnel and 3,000 tons of relief materiel. Il-76s from the 13th Transport Division reportedly carried 55 percent of air-transported freight and 70 percent of air-dropped supplies for the operations.[91]

Overall, however, the PLAAF's performance faced considerable criticism. The former Chengdu MRAF, initially in charge of coordinating air-based relief, relied on the 4th Transport Division, which consisted of a few medium-sized helicopters for search and rescue and other aging platforms. The other military regions contributed over 100 medium-sized helicopters as well as Y-8 transport planes. The lack of heavy-lift helicopters severely limited the ability of the PLA to carry out disaster relief in a timely manner. Because of its inadequate capacity, the PLAAF had to rely on civilian assets and contributions from PLAN remote-sensing aircraft to assist with reconnaissance.[92] Moreover, inadequate training led pilots to drop materiel from

---

[87] "Il-76s in a Hurry: Taking Stock of Major Operations by Chinese Air Force Il-76 Transport Aircraft" [伊尔很忙: 盘点中国空军伊尔-76运输机重大行动], Sina Photographs, December 9, 2014.

[88] "PLA Details Chinese Military Operations Other than War Since 2008," *PLA Daily* in English, September 6, 2011.

[89] "PLA Details Chinese Military Operations Other than War Since 2008," 2011.

[90] Li Guowen [李国文], Wu Dechao [吴德超], Zhu Bin [朱斌], Zheng Wei [郑蔚], and Qian Bei [钱蓓], "Decoding the People's Air Force's Core Military Capabilities" ["解读人民空军核心军事能力"], *Wen Hui News* 《文汇报》, November 2, 2009. A version of this article ran in *Air Force News* 《空军报》 on November 11, 2009, p. 3.

[91] "Il-76s in a Hurry" ["伊尔很忙"], 2014.

[92] "China Initiates First Class Emergency Response for Quake Relief in Qinghai," Xinhua [新华] in English, April 14, 2010.

higher elevations, leading to inaccurate drops. Twice, poorly trained pilots failed to land a helicopter in the epicenter of the disaster zone.[93]

Since 2002, the PLAAF has carried out humanitarian assistance missions internationally, usually in response to natural disasters in the region. Table 3.3 summarizes the PLAAF's known participation in HA/DR efforts outside of China.

Table 3.3. Known PLAAF International HA/DR Operations

Year	Country	Type of Crisis	PLAAF Units Involved	Materials Delivered (Tonnage and Value of Supplies, If Known)
2002	Afghanistan	War	At least two Il-76s	Humanitarian supplies
2010	Mongolia	Winter storm	Three Il-76s	Emergency food supplies, blankets, generators
2010	Pakistan	Flooding	Three Il-76s	Medical kits, tents, generators, water purification equipment (69 tons, 10 million renminbi (RMB))
2011	Pakistan	Flooding	At least four Il-76s	Tents (estimated 400 tons, 30 million RMB)
2011	Thailand	Flooding	At least three Il-76s	Small boats, life rafts, generators, emergency lamps, water-purifying equipment, sandbags, solar-powered flashlights (72+ tons)
2014	Maldives	Water shortage	Two Il-76s	Drinking water (40 tons)
2015	Malaysia	Flooding	Two Il-76s	Water-purifying equipment, sewage pumps
2015	Sri Lanka	Flooding	Not known	Blankets, towels, powdered milk, sugar (72 tons, 10 million RMB)
2015	Nepal	Earthquake	At least four Il-76s	PLA medical personnel; medical equipment; humanitarian supplies including tents, blankets, generators (416 tons)
2015	Myanmar	Flooding	Two Il-76s	Blankets, tent lanterns, generators (10 million RMB)

The PLAAF's first known operation of this type occurred in 2002, although it appears to have been largely symbolic. While operating as "China United Airlines" aircraft in 2002, but with military flight crews, at least two Il-76s brought humanitarian supplies to Kabul Airport via seven sorties.[94]

---

[93] Nirav Patel, "Chinese Disaster Relief Operations: Identifying Critical Capability Gaps," *Joint Forces Quarterly*, No. 52, first quarter 2009, pp. 111–117. The PLAAF also carried out HA/DR activities following the 2010 Yushu earthquake. On August 4, 2014, a regiment (likely from the former Chengdu MRAF's 4th Transport Division) also sent a Y-8 to participate in relief efforts following an earthquake in Ludian County, Yunnan Province.

[94] "Il-76s in a Hurry" ["伊尔很忙"], 2014. Sina.com website, "Il-76 Transport Aircraft Large-Scale Activity," December 9, 2014. This operation is also alluded to in a 2012 article on the 13th Transport Division: "Qiu Defu [邱德甫] and his crew flew to Afghanistan in 2002 to deliver relief materials. At that time, Afghanistan belonged to the first-type of region in the chaos of war, areas of chaos caused by war, and all flights inside the country were subject to the rules of control in a wartime air zone managed by the unified management of the U.S.-led multinational forces

Chinese HA/DR operations increased significantly beginning in 2010, perhaps reflecting lessons learned and greater confidence gained following the domestic relief mission in Wenchuan. Following a blizzard and sudden cold snap in Mongolia in early 2010, the PLAAF transported relief supplies to the country.[95] After flooding in Pakistan during summer 2010, China sent financial assistance to Pakistan.[96] The PLAAF also participated in the relief efforts, with three Il-76s transporting supplies to Pakistan.[97]

After floods hit Pakistan again in September 2011, four Il-76s transported 7,000 waterproof tents to the afflicted areas.[98] In response to flooding in Thailand one month later, the PLAAF transported two rounds of supplies.[99] One report stated that China sent five aircraft in total to Bangkok, at least one of which was a chartered flight.[100]

The PLAAF's role in HA/DR efforts grew more frequent in 2014 and 2015. In December 2014, two of the PLAAF's Il-76s delivered drinking water to the Maldives when the island suffered an acute water shortage.[101] In January 2015, the PLAAF sent two Il-76s loaded with equipment to Malaysia following flooding there.[102] A few weeks later, China transported humanitarian aid to flood-stricken Sri Lanka.[103]

---

according to the rules of control in the air zones of war, and the flight conditions were very harsh." See Dong Ruifeng [董瑞丰] and Guo Hongbo [郭洪波], 2012.

[95] "Facts and Figures: Chinese Air Force Observes 65th Anniversary," Xinhua [新华] in English, November 4, 2014.

[96] "China Pledges $250m Flood Aid to Pakistan," *China Daily* in English, December 18, 2010.

[97] Guo Hongbo [郭洪波], "Three Chinese Air Force Transport Aircraft Carry Relief Provisions to Pakistan" ["中国空军 3 架运输机向巴基斯坦运送救灾物资"], *China News Service* 《中国新闻网》, August 4, 2010.

[98] On the first trip to Pakistan, they delivered 3,000 tents with a volume of 460 cubic meters and weighing 170 tons. After unloading the supplies, the four Il-76s returned home and with the help of an unknown fifth aircraft, transported the remaining supplies to Pakistan. "Our Country Deploys Il-76s to Pakistan to Deliver Relief Supplies" ["我空军出动伊尔 76 运输机向巴基斯坦运送救援物资"], *China Online* 《中国网》, September 23, 2011.

[99] "Chinese Air Force Sends Three Transport Aircraft to Deliver Relief Supplies for Thailand Flood," *People's Daily Online* 《人民网》, October 22, 2011; Kenneth Allen and Emma Kelly, "Assessing the PLAAF's Foreign Relations," *China Brief*, Vol. 12, No. 9, April 26, 2012; "Chinese Government's Third Batch of Relief Supplies to Thailand Arrives in Bangkok" ["中国政府向泰国提供的第三批援助物资运抵曼谷"], Central Government Web Portal [中央政府门户网站], October 22, 2011.

[100] "Chinese Government's Third Batch of Relief Supplies to Thailand Arrives in Bangkok" ["中国政府向泰国提供的第三批援助物资运抵曼谷"], 2011.

[101] "Il-76s in a Hurry" ["伊尔很忙"], 2014.

[102] "Chinese Military Delivers Aid to Malaysia," Xinhua [新华] in English, January 12, 2015.

[103] It is implied that the PLAAF used airlift assets to move the supplies, given the discussion of "transport responsibilities" and the short timeframe to deliver the supplies. Zhang Ziyang [张子扬], "Chinese Armed Forces Ship Ten Million RMB Worth of Aid Materials to Sri Lanka" ["中国军队启运 1000 万元援助斯里兰卡物资"], *China News Service* 《中国新闻网》, January 25, 2011.

Following the earthquake in Nepal in 2015, an Il-76 transported 37 medical personnel as well as medical equipment and disaster relief provisions.[104] Additional Il-76s loaded with humanitarian supplies arrived in Kathmandu in the days following the earthquake.[105] A few months later in August, the PLAAF sent two transport aircraft to flood-stricken Myanmar.[106]

Overall, the PLAAF's international HA/DR operations to date have been characterized by a few common details. In most operations, two to four Il-76s from the former Guangzhou MRAF's 13th Transport Division have carried out the mission, assembling at major Chinese airports (including commercial airports) to load supplies and then transporting them to the country in need, often to its largest or capitol airport. From accounts of these operations, it does not appear that the PLAAF has helped transport supplies from the major airports in these countries to the specific areas afflicted. Supplies have largely consisted of foodstuffs, water, temporary housing and power generation supplies, and small boats in the case of flooded areas. Although participation in international HA/DR represents an important step forward for the PLAAF, it is important to note that these operations have been limited in scale and scope compared with a number of HA/DR efforts by the United States, as well as day-to-day transport missions flown by U.S. aircraft. For comparison, one USAF publication notes that, in 2013, an average of one USAF transport aircraft took off every 90 seconds to deliver personnel and cargo to locations around the world.[107]

Noncombatant Evacuation Operations

The large-scale evacuation of Chinese nationals from Libya in February–March 2011 marked China's first noncombatant evacuation operation outside China that involved military aircraft. Four Il-76 transport aircraft from the Guangzhou MRAF 13th Transport Division were dispatched to Libya with refueling stops in Karachi, Pakistan, and Khartoum, Sudan.[108] PLAAF aircraft participating in the NEO reportedly "flew across five countries, across the Arabian Sea and the Red Sea, across six time zones . . . over more than 9,500 kilometers."[109] Once in the

---

[104] "Chinese Air Force's First Assistance Aircraft Arrives in Nepal Earthquake Disaster Area" ["中国空军首架救援飞机飞抵尼泊尔地震灾区"], *China Military Online* 《中国军网》, April 28, 2017.

[105] "Chinese Air Force's First Assistance Aircraft Arrives in Nepal Earthquake Disaster Area" ["中国空军首架救援飞机飞抵尼泊尔地震灾区"], April 28, 2017; "Air Force Planes Transport Materiel to Nepal," *China Daily Online* in English, April 29, 2015; "China Sends Record Military Personnel Numbers to Nepal," Xinhua [新华] in English, May 7, 2015. Chinese-language descriptions of the relief effort suggest that four Il-76s were sent in total, while English-language reports state that "eight transport planes" were deployed.

[106] "Love Without Limits, Surge of Compatriot Feelings! Chinese Government's Myanmar Disaster Relief Assistance Activities; Temporary Housing Handover Ceremony Held in Yangon Harbor" ["爱无界 胞波情！ 中国政府援助缅甸活动板房交接仪式在仰光港举行"], *Myanmar Golden Phoenix* 《金凤凰》, March 11, 2016.

[107] U.S. Air Force, "Global Vigilance, Global Reach, Global Power for America," August 2013, p. 7.

[108] Tan Jie, "PLA Air Force Transporters Evacuate Compatriots from Libya," *China Military Online* in English, March 2, 2011.

[109] Dong Ruifeng [董瑞丰] and Guo Hongbo [郭洪波], 2012 .

region, the four Il-76s ferried two rounds of evacuees from Sebha Airport in Libya to Khartoum International Airport for a total of eight sorties; Chinese media stated that each plane flew over 30,000 kilometers in 46 hours.[110] By the evening of March 2, the Il-76s had moved 1,655 Chinese citizens from Libya to Khartoum.[111] They also transported 287 Chinese citizens directly to Beijing Nanyuan Airport.[112]

According to Chinese sources, the PLAAF was sent because "the number of people to be evacuated was large, and some foreign airliners that had been leased broke the agreed arrangement and became unavailable."[113] Moreover, use of military aircraft had additional symbolic import because they signified a "state action, directly representing the state will."[114] However, based on the PLAAF's relatively limited role in the operation, its involvement seems to have been either symbolic or possibly an initial test of transport units' capabilities to conduct overseas operations in the event of a crisis. As noted above, the PLAAF brought 1,655 Chinese citizens from Libya to Sudan and transported 287 Chinese evacuees from the region back to China. Even assuming no overlap in those two groups of evacuees, the PLAAF transported, at most, less than six percent of total Chinese citizens evacuated out of Libya (35,680).[115]

The PLAAF has not played a role in subsequent evacuations of Chinese nationals. Kenneth Allen and Phillip Saunders note that, during the Ebola crisis in November 2014, rather than sending PLAAF transport aircraft, China used nine civilian aircraft from China Eastern Airlines and China Cargo Airlines to transport 282 medical staff and 767 tons of medical materials to Guinea, Liberia, and Sierra Leone.[116] The PLAAF also did not participate in the PLAN's evacuation of civilians in Yemen in 2015. Chinese civilian aircraft, with the assistance of Egyptian civilian aircraft, flew Chinese citizens home from Djibouti after the PLAN evacuated them from Yemen to Djibouti.[117]

---

[110] Zhang Jinyu and Shen Jinke, "PLA Air Force Transporters Bring Home Chinese Evacuees from Libya," *China Military Online* in English, March 4, 2011; "PLA Details Chinese Military Operations Other than War Since 2008," 2011.

[111] Zhang Jinyu and Shen Jinke, 2011.

[112] "Il-76s in a Hurry" ["伊尔很忙"], 2014.

[113] Dong Ruifeng [董瑞丰] and Guo Hongbo [郭洪波], 2012.

[114] Dong Ruifeng [董瑞丰] and Guo Hongbo [郭洪波], 2012.

[115] Zhang and Shen, 2011; "Military Aircraft Bring Back 287 Chinese from Libya," Xinhua [新华] in English, March 5, 2011; Bo Xu, 2011, p. 47. The Il-76s reportedly transported at least 240 foreigners out of Libya as well: "Chinese Air Force Evacuates 751 from Libya," Xinhua [新华] in English, March 2, 2011.

[116] See Xiang Bo, "Nine Chinese Aircraft Finish Ebola Rescue Task," Xinhua [新华], November 19, 2014. From Kenneth Allen, Phillip C. Saunders, and John Chen, "Chinese Military Diplomacy, 2003–2016: Trends and Implications," *China Strategic Perspectives* 11, Institute for National Strategic Studies, National Defense University, July 2017, p. 42.

[117] "Chinese Ambassador to Djibouti States Yemen Evacuation Work Proceeded Smoothly" ["中国驻吉布提大使说也门撤侨工作顺利进行"], *People's Daily* 《人民网》, April 1, 2015.

Personnel Recovery

PLAAF aircraft have assisted in the recovery of remains of deceased nationals. In 2004, the 13th Transport Division sent an airplane to Afghanistan to bring home 11 nationals who died in Afghanistan.[118] In March 2016, an Il-76 flew to South Korea to retrieve remains of Chinese soldiers who were killed during the Korean War.[119]

PLAAF aircraft also participated in the search for missing passengers from the Malaysian Airlines Flight 370 in March 2014. The PLAAF dispatched two Il-76 aircraft and one Y-8 to help conduct searches of the southern Indian Ocean.[120]

At least one Il-76 operated from Perth International Airport to conduct visual searches for the aircraft wreckage through April, completing at least 16 search missions.[121] Reports claimed that each maintenance crew worked at least 14 hours per day to support the deployment.[122] One media report acknowledged substantial hurdles to carrying out the search, including the vast distance and poor weather.[123] The lack of specialized equipment designed to spot missing personnel further limited the utility of the Il-76s for the personnel recovery mission. Although these types of aircraft might thus seem to be unusual choices for this mission, their inclusion likely reflected the importance China attached to the mission. In all, the PLAAF's decision to send the Il-76s underscores the limited ability of the PLAAF to operate outside its borders. Although it has search-and-rescue aircraft, these lack the range and experience to fly at such distances. The PLAAF remains dependent on a small number of large transports to carry out the great majority of its overseas missions, even when the planes are ill-suited to the task.

This may explain why, in recent years, the PLA has turned to two other types of aircraft for specialized, small-scale missions abroad. In June 2016, the Transportation Bureau [运输投送局] under the CMC Logistics Support Department (LSD) signed an agreement with the Red Cross Society of China to use civil-military air ambulances "to implement transportation and transfer of the sick and wounded of the PLA."[124] One of the aircraft, operated by Beijing Airlines and painted with the Red Cross symbol and the phrase "China Aviation Medical Assistance" ["中国航空医疗救援"], traveled to South Sudan in July 2016 to transport two injured peacekeepers back to China, arriving back in China after an 18-hour nonstop flight. Based on photos of the

---

[118] "Chinese Air Force to Fly Tomorrow to Afghanistan to Receive the Remains of Victims" ["中国空军专机明日赴阿富汗接遇难者遗体"], Southcn.com [南方网], June 13, 2004.

[119] China Military Online, "China Aircraft Arrives in ROK to Return Martyrs' Remains," *China Military Online* in English, March 29, 2016.

[120] "Il-76s in a Hurry" ["伊尔很忙"], 2014.

[121] "Record of Chinese Air Force's Search for MH370," *China Military Online* in English, April 11, 2014.

[122] "Record of Chinese Air Force's Search for MH370," 2014.

[123] "PLAAF Overcomes Disadvantages to Commit Full Efforts to Searching Missing Plane," *China Military Online* in English, March 28, 2014.

[124] "China's First Military Air Ambulance Debuts," *China Military Online* in English, August 22, 2016.

flight, at least one member of the PLA flew with the injured peacekeepers aboard the plane.[125] One article cited an official from the LSD Transportation Bureau remarking that as PLA operations overseas become more frequent, "the PLA should rely on the advantages of regional [civilian] aviation rescue institutions to provide an air to ground model that provides direct emergency aviation transport and transfer services for the armed forces' sick and wounded."[126] According to the same article, the air ambulance program may be integrated into the PLA's strategic reserve forces.

In 2016, PLAAF civilian personnel–transport aircraft, commonly used for VIP travel, were also sent overseas to transport the remains of Chinese peacekeepers killed in Mali and South Sudan. In June 2016, a PLAAF-owned Boeing 737 flew to Mali carrying a PLA working group headed by Major General Su Guanghui, the acting director of the Ministry of National Defense's Peacekeeping Affairs Office.[127] The plane returned carrying the remains of one peacekeeper. In July 2016, the same PLAAF aircraft was again dispatched with a working group headed by Major General Su, this time to retrieve the remains of two PLA soldiers killed in a United Nations peacekeeping operation in South Sudan.[128] The aircraft was modified to accommodate stretchers for two other peacekeepers injured during the same operation.[129]

Demonstrations (Competitions and Air Shows)

In recent years, the PLAAF has increased its participation in competitions and air shows abroad. The PLAAF has sent fighter aircraft to participate in the international portion of the Russia-organized Aviadarts competition in 2014, 2015, and 2016. For the 2014 event, competition areas included navigation, reconnaissance, aerobatics, and air-to-ground attack.[130]

---

[125] "Injured Peacekeepers Return Home from South Sudan," Xinhua [新华] in English, July 17, 2016; "China's First Military Air Ambulance Debuts," 2016.

[126] "China's First Military Air Ambulance Debuts," 2016; "Our Military Expands Civil-Military Fusion to News Roads of Aviation Emergency Response Assistance" ["我军拓展军民融合航空应急救援新路"], *PLA Daily* 《解放军报》, August 23, 2016.

[127] "Our Military Working Group Goes to Mali to Launch Efforts Followed Ambushed Peacekeeping Personnel Casualties; Conduct Rescue and Medical Treatment Work" ["我军工作组赴马里开展遇袭伤亡维和人员善后和救治工作"], *China Military Online* 《中国军网》, June 2, 2016; "Chinese Military Team Starts Working in Gao of Mali," *China Military Online* in English, June 6, 2016; "Body of Chinese Peacekeeper Being Flown Back to China," *CRI English*, June 8, 2016; "Chinese Peacekeeper's Body Brought Home from Mali," Xinhua [新华] in English, June 9, 2016; Zheng Jinran, "Peacekeeper's Body Returns for Burial," *China Daily* in English, June 10, 2016; "Plane Carrying Body of Deceased Chinese Peacekeeper Has Arrived in Changchun," CCTV.com, June 13, 2016.

[128] The tail number on the plane, B-4018, is the same in photos of both incidents. For more information, see "B-4018 PLAAF—China Air Force Boeing 737-33A," Planespotters.net, 2016; "Remains of Chinese Peacekeepers Brought Home," *China Military Online* in English, July 20, 2016; Luo Zheng [罗铮] and Zhu Hongliang [朱鸿亮], "Chinese Military Working Group Takes Military Aircraft and Rushes to South Sudan" [中国军队工作组乘军机赶赴南苏丹], *China Military Online* 《中国军网》, July 13, 2016.

[129] "Remains of Chinese Peacekeepers Brought Home," 2016.

[130] "Chinese Air Force Attends 'Aviadarts 2014' International Pilot Competition in Russia," *China Military Online* in English, July 21, 2014.

To provide a sense of the scope of competition, the two competing Chinese Su-30 aircraft reportedly launched 24 rocket projectiles and fired 60 cannon rounds. For the ground-attack section, the competing aircraft used unguided ordnance per competition guidelines.[131]

In 2015, China sent JH-7 fighter-bombers to Aviadarts. A *PLA Daily* article explained that Russia provided ammunition for the 2014 exercise because all of the participants flew Russian-made fighters during the competition. Because the PLAAF brought Chinese-made aircraft for the 2015 exercise, however, Russia allowed China to transport appropriate ammunition supplies via transport aircraft.[132] Although the PLAAF placed second in the 2014 competition (following Russia and ahead of Belarus), the PLAAF placed third out of at least seven participants in Aviadarts 2015.[133]

In Aviadarts 2016, two of the PLAAF's JH-7A fighter-bomber aircraft competed against Russia, Kazakhstan, and Belarus.[134] The aircraft came from an unidentified air brigade in Nanjiang (Western TCAF).[135] An Il-76 transport aircraft also traveled to Russia with a military reporter on board.[136] A photo of the opening ceremony shows roughly 30 to 40 PLAAF personnel in attendance.[137] In the competition, the JH-7As focused on the accuracy of attacking ground targets with air-to-surface rockets and bombs, placing second overall.[138]

One article noted that Aviadarts 2016 revealed Chinese pilots' weakness in conducted visual reconnaissance compared with their Russian counterparts. The commander of the participating Chinese units, Western TCAF aviation brigade commander Shen Yuanzhong [沈元中] stated that "some of the basic skills of pilots, such as visual reconnaissance, have degenerated."[139] He blamed excessive "dependence on advanced technology," commenting that his brigade intended

---

[131] "'Aviadarts 2014' International Pilot Competition Ends," *China Military Online* in English, July 29, 2014.

[132] "Chinese Military Transport Planes Carry Ammunitions to Russia for Int'l Games," *China Military Online* in English, July 30, 2015.

[133] "China Ranks Second in 10 Team Competitions at International Military Games 2015," *China Military Online* in English, August 17, 2015.

[134] It appears that three were sent to Russia but only two participated. See "China's JH-7 Wins Runner-Up of 'Aviadarts-2016' Competition in Russia," *China Military Online*, August 9, 2016; "JH-7 Fighter Bombers Participate in International Games in Russia," *China Military Online*, August 4, 2016.

[135] Identified based on the personnel mentioned at the event. See *Directory of PRC Military Personalities*, Washington, D.C.: Defense Intelligence Agency, 2016, p. 95.

[136] "China to Participate in International Army Games 2016," *China Military Online* in English, July 1, 2016. The reporter can be seen in a news clip: Wang Xiaoyu [王晓玉], Yan Qian [闫倩], and Zhou Limin [周立民], "Air Force Combat Aircraft Participating in 'International Military Competition–2016' Arrive in Russia" ["空军参加 '国际军事比赛-2016' 战机抵俄"], CCTV-7 "Military Report" [军事报道] program, July 26, 2016.

[137] "'Aviadarts' International Contest Kicks Off," *China Military Online* in English, August 16, 2016.

[138] "'Golden Darts' Comes to an End; Chinese Air Force Participants Seize Second Place" ["'航空飞镖'闭幕 中国空军参赛队夺第二名"], *China Military Online* 《中国军网》, August 10, 2016; "China's JH-7 Wins Runner-Up of 'Aviadarts-2016' Competition in Russia," 2016.

[139] "Chinese Airmen Hone Skills in International Competition," Xinhua [新华] in English, August 28, 2016.

to address these deficiencies by conducting more exercises and drills with live ammunition and "real ground targets."[140]

The Bayi Aerobatics Team has performed in two international air shows beyond China's borders and has visited three countries so far. In August 2013, seven J-10s and two Il-76s participated in the 2013 Moscow Air Show.[141] In 2015, the team performed at the Langkawi International Maritime and Aerospace Exhibition in Malaysia.[142] After performing in the air show, the Bayi Aerobatics Team flew back through Thailand and participated in a military exchange program.[143] The Bayi Aerobatics Team additionally returned to Thailand following the Falcon Strike exercise in November 2015.[144]

Challenges in Conducting Overseas Operations

Owing to its relative inexperience with overseas operations, the PLAAF has faced diverse challenges. Some arise from the nature of operating in unfamiliar harsh environments, while others stem from maintenance and logistics problems related to the high operational tempo of scarce assets, especially large transport planes. Chinese media have also noted challenges in navigating over vast distances and hurdles arising from international diplomacy regarding use of airspace or from working with personnel from different cultures. The need to bring all munitions, maintenance, supply, communications, and other equipment required to carry out air operations has imposed a major burden, especially for any activity that involves more than one or two aircraft or that feature dissimilar aircraft in complex training and other operations.

Sustainment and Prepositioning

Fuel; Petroleum, Oil, and Lubricants; and Logistics

The PLAAF has worked for years on sustainment issues in harsh climates within the country, especially in the mountainous western regions. A 2012 article described efforts to overcome logistics bottlenecks to extended deployments in high-altitude locations, such as extending the shelf life of petroleum, oil, and lubricants (POL), and to improve facilities and oxygen access to

---

[140] "Chinese Airmen Hone Skills in International Competition," 2016.

[141] "Ten Highlights of China's Military Diplomacy in 2013," *PLA Daily* in English, December 23, 2013.

[142] "PLAAF Aerobatic Team Arrives in Malaysia for Stunt Shows," *China Military Online* in English, March 12, 2015.

[143] "Direct Hit China-Thai Air Forces' First Joint Flight Exhibition" ["直击中泰空军首次联合飞行表演"], *China Daily Online* 《中国日报中文网》, November 30, 2015.

[144] Patpicha Tanakasempipat and Jutarat Skulpichetrat, "China, Thailand Joint Air Force Exercise Highlights Warming Ties," Reuters, November 24, 2015.

pilots. The article stated that experimentation and testing resulted in the development of POL products better suited to the harsh conditions, thereby facilitating PLAAF operations.[145]

Spare Parts and Maintenance

Heavy use of the limited supply of transport planes has exacerbated challenges in maintaining the aging platforms. PLAAF media acknowledge problems of mechanical problems. During one mid-March 2015 exercise, an article reported that heavy use of one regiment's aircraft had made "accidental malfunctions relatively frequent."[146]

Another article acknowledged shortcomings in the maintenance of the 13th Transport Division's Il-76s. It mentioned insufficient technical information about equipment and instruments aboard the aircraft, a lack of maintenance equipment, and insufficient numbers of skilled maintenance personnel.[147]

Limited Large-Transport Capacity

The limited availability of heavy transports remains a major constraint on the PLAAF's ability to deploy aircraft in an expeditionary manner. To overcome this limitation, the service has expanded its use of civilian cargo planes. For example, during the Stride 2009 two-month exercise, PLA General Headquarters "coordinated the mobilization and requisitioning of quite a few civilian passenger and cargo aircraft from relevant aviation companies."[148] A media report stated that this was a "first in the history of PLA exercises to feature the levying of civilian transport aircraft to move troops across long distances."[149]

In some cases, military authorities have mobilized civilian assets to assist with the transportation of troops and equipment abroad. In Mission Action 2013, media reports stated that "two lightly-equipped units participating in the exercise flew to the assembly area by taking civil and military transport aircrafts from the Changle Airport and a military airfield in Fuzhou

---

[145] Li Qiang [李强] and Yin Jiahua [尹家骅], "Air Force Logistics Department Organizes Special Topic Seminar on Accelerating the Building of Comprehensive Support Capabilities" ["空军后勤部组织专题研讨 加快推进部队高原常态化驻训综合保障能力建设"], *Air Force News* 《空军报》, June 7, 2012, p. 1.

[146] Liu Gang [刘纲] and Yu Yi [余毅], "Establish Scientific Mechanisms, Oversee Each Phase: Airborne Air Transport Regiment Training and Tasks Progress Steadily" ["建立科学机制 严把每个环节: 空降某航运团训练和任务同向稳步推进"], *Air Force News* 《空军报》, March 26, 2015, p. 1.

[147] Yang Yongsheng [杨永生], Zhu Zhanghu [朱章虎], and Zhao Lingyu [赵凌芋], 2012.

[148] "Jinan Military Region Requisitions Civilian Aircraft for Long-Distance Transfer of Exercise Troops" ["济南军区参演部队征用民航飞机远程投送兵力"], Xinhua [新华], August 17, 2009.

[149] "Jinan Military Region Requisitions Civilian Aircraft for Long-Distance Transfer of Exercise Troops" ["济南军区参演部队征用民航飞机远程投送兵力"], 2009.

respectively."[150] Four Boeing 737-800s helped carry "rifles, sniper rifles, rocket propelled grenades, and other light firearms as well as all the armaments of individual soldiers."[151]

Long-Distance Navigation and Instrumentation

Articles related to the 13th Transport Division's operations acknowledge challenges related to long-distance navigation and instrument controls to enable greater autonomy. According to one article, Il-76 crews initially struggled to maintain bearing over extremely long distances. Because of a lack of specialized equipment, the personnel had trouble controlling drift in flight. Technicians had to improvise solutions with new instruments and equipment to correct the problem. The article referenced the "upgrading and conversion of command monitoring facilities—including primary and secondary radar and meteorological signals, shortwave radio, ultra-shortwave radio, and other command and monitoring equipment."[152] These changes not only improved navigation and command and control, they also enabled crews to have real-time information about weather conditions at hundreds of "domestic and foreign airfields." The units also retrofitted equipment such as ultra-shortwave radio, collision avoidance systems, and improved independent support capabilities, monitoring, and flight clearance to enable airplanes to cope with risks autonomously.[153]

Command and Control of Dissimilar Aircraft Operations

The challenges of operating a single large transport may be imposing enough for the relatively inexperienced PLAAF, but the difficulties are even greater for larger groupings of dissimilar aircraft, especially when performing complex operations related to combat training. For the Peace Mission 2007 exercise that took place in Russia, a signals regiment prepared and trained for 84 days to execute the deployment. This was the regiment's first time traveling outside the country to provide support for an exercise, and the regiment was required to carry along "more than a hundred sets of equipment," of which one report claimed 30 percent was "new."[154] Reports claimed that participants found the deployment challenging, because of a lack of familiarity with the weather, the terrain, or the electromagnetic environment. Another challenge was that the personnel quota to work the equipment was "half of the 100 people

---

[150] "Mission Action–2013 Cross-Regional Maneuver Exercise: Military-Civilian Air Transport Forces Jointly Deliver Lightly-Equipped Units" ["'使命行动-2013' 跨区机动演习: 军地空中运力联合投送轻装部队"], CCTV-13 News "Live News" [新闻直播间] program, September 15, 2013.

[151] "Mission Action–2013 Cross-Regional Maneuver Exercise," 2013.

[152] Yang Yongsheng [杨永生], Zhu Zhanghu [朱章虎], and Zhao Lingyu [赵凌芋], 2012.

[153] Yang Yongsheng [杨永生], Zhu Zhanghu [朱章虎], and Zhao Lingyu [赵凌芋], 2012.

[154] Zhang Changsheng [张长胜], Ma Junhong [马军鸿], and Li Benqian [李本钱], "Air Force Signals Regiment Participates in Joint Military Exercise Peace Mission 2007" ["空军某通信团参加 '和平使命-2007' 联合军演纪事"], *Air Force News* 《空军报》 September 18, 2007, p. 1.

usually allocated to the equipment."[155] The team carried out cross-training to ensure each team member could carry out multiple tasks. An example cited was a ground-to-air communications operator who also trained to carry out vehicle maintenance, satellite communications, military mobile communications, and who could also serve as a driver. Noncommissioned officers grades 5 and higher were expected to master six types of equipment, while more junior members were to operate three or more types.

A sense of the scope of the support equipment required to carry out the mission may be gleaned from reports that the team carried "thousands of items" and "35 kilometers of communications and electrical cable." The team encountered unexpected difficulties using Russian electrical power supplies and had to adapt their equipment accordingly. Reports also described how the team established temporary command systems. In Russia, the team set up communications at Shagol Airfield and at the live forces exercise area for ground-to-air communications.[156]

Access and Diplomacy

Media reports do not generally provide much information on issues related to access and diplomacy. However, occasional insights may be discerned from disparate reports. An article on the recovery of the remains of 11 Chinese nationals in Afghanistan mentioned that the Ministry of Foreign Affairs assisted with making arrangements to ensure access to Afghanistan.[157] Articles about the 2011 Libya NEO mentioned that, in addition to the Chinese defense attaché to Libya and his secretary [秘书], the attachés and secretaries from Tunisia and Greece helped coordinate the overall evacuation efforts, including PLAN and civilian aircraft flights.[158]

Crews flying aircraft associated with the 13th Transport Division have discussed their learning experiences regarding operating in the air space of other countries. In one article, crew members reported that an unnamed Southeast Asian country did not allow the PLAAF aircraft to fly across its territorial airspace. The crew then had to choose a different route. The article concluded that the PLAAF needed to "know more about international rules, local conditions and customs of relevant foreign countries, and how foreign militaries carry out flight training and management."[159]

---

[155] Zhang Changsheng [张长胜], Ma Junhong [马军鸿], and Li Benqian [李本钱], 2007.

[156] Zhang Changsheng [张长胜], Ma Junhong [马军鸿], and Li Benqian [李本钱], 2007.

[157] "Chinese Air Force to Fly Tomorrow to Afghanistan to Receive the Remains of Victims" ["中国空军专机明日赴阿富汗接遇难者遗体"], 2004.

[158] "Chinese Defense Attachés Participate in All-Out Task to Evacuate Chinese Personnel in Libya" ["中国武官全力参与在利比亚中方人员撤离工作"], Xinhua [新华], February 28, 2011.

[159] Dong Ruifeng [董瑞丰] and Guo Hongbo [郭洪波], 2012.

Conclusion

PLAAF units have been conducting long-distance deployments within China since at least the late 1990s and early 2000s. In recent years, the Communist Party of China has directed the PLA to develop capabilities to protect China's interests abroad—both regionally and globally. The PLAAF has made incremental progress in its ability to carry out overseas operations. The added complexity of carrying out nonwar missions exposed the limitations of the PLAAF in the mid-2000s. Domestic experiences, such as participation in the 2008 Sichuan earthquake rescue effort, helped the PLAAF improve its abilities to navigate across vast distances, exercise command and control, and anticipate logistics and maintenance needs at remote locations. The PLAAF has also learned valuable lessons regarding operations abroad in subsequent training, relief, and other real world nonwar missions. The first international exercise involving PLAAF participation overseas was Peace Mission 2007. The PLAAF has also carried out relief efforts to other countries as well as its first NEO involving the use of military aircraft to evacuate Chinese citizens from Libya in 2011. In addition, the PLAAF has recently expanded its role in foreign military exchanges by attending international air shows and air competitions beginning in 2013 and 2014, respectively. Figure 3.1 summarizes these deployments abroad from January 2002 through October 2016. Through these experiences, small numbers of Chinese pilots are learning to navigate abroad, manage issues of diplomatic access, and operate with greater autonomy.

Figure 3.1. PLAAF International Deployments Are Growing

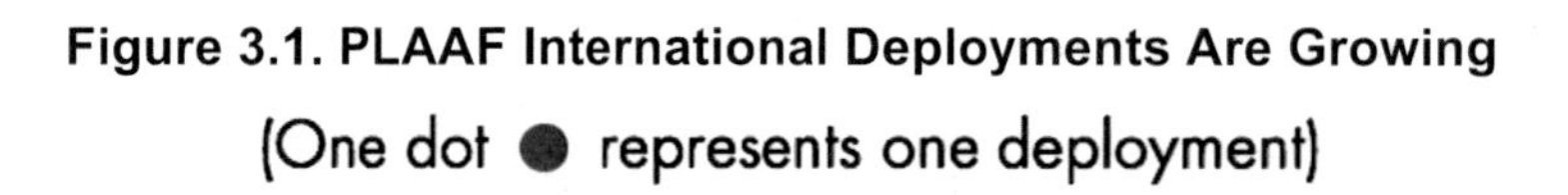

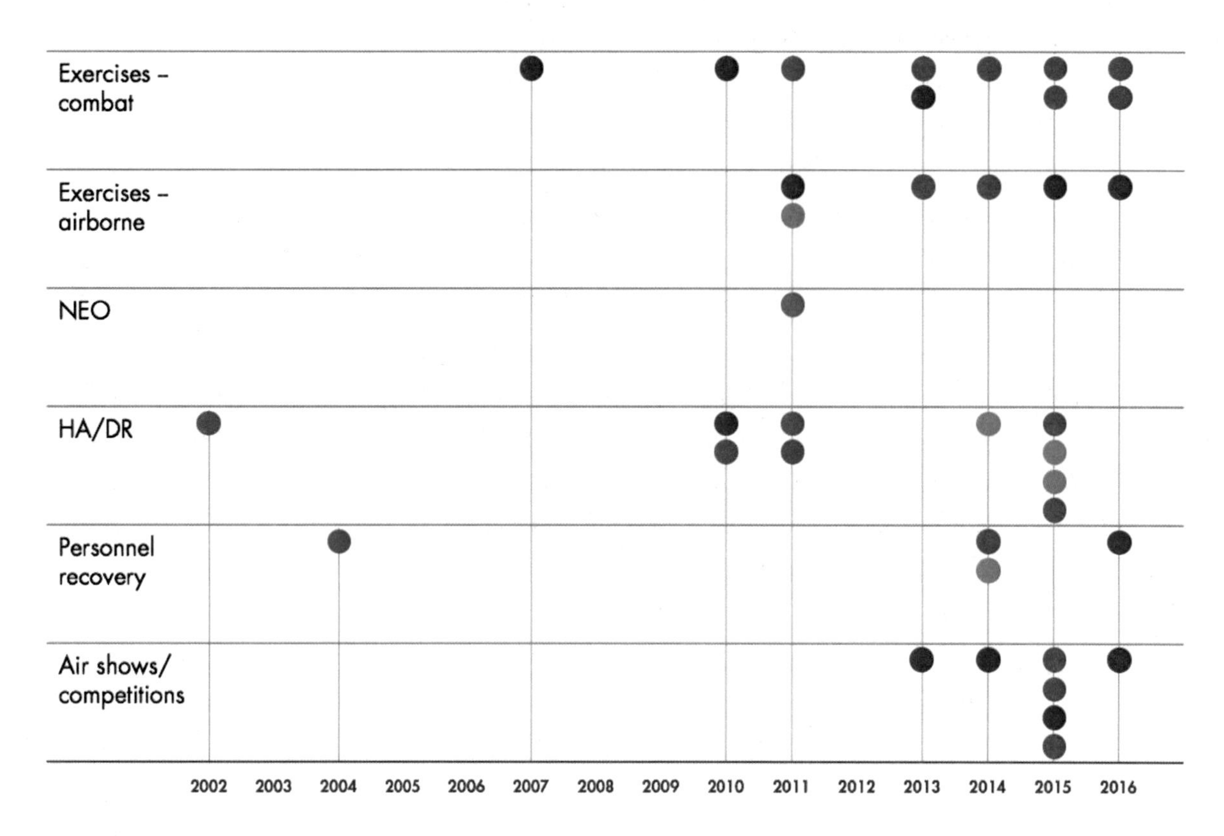

- **Af-Pak** Afghanistan, Pakistan
- **Central/West Asia** Belarus, Kazakhstan, Russia, Turkey
- **Northeast Asia** Mongolia, South Korea
- **Latin America** Venezuela
- **Southeast Asia** Brunei, Indonesia, Malaysia, Myanmar, Thailand
- **South Asia/Oceanic** Australia, Maldives, Nepal, Sri Lanka
- **Africa** Libya

NOTE: Includes known operations between January 2002 and October 2016.
Does not include operations by PLAAF civilian transport aircraft.

To improve its expeditionary capabilities, the PLAAF has focused heavily on developing a small number of reliable, elite units to carry out politically high-profile missions abroad. The most important of these units is the Il-76-equipped 39th Regiment of the 13th Transport Division. While this approach suits the PLAAF's limited resources and incomplete modernization, it also reveals that the PLAAF continues to face serious constraints on its ability to operate abroad. Among the most important are a scarcity of high-demand large-transport aircraft and a lack of overseas infrastructure to support sustained operations. The aging Il-76 aircraft of the 13th Transport Division remain responsible for carrying out the great majority of the PLAAF's expeditionary activity. China has fewer than two dozen of these aircraft, however, and media reports acknowledge that the high frequency of use has stressed their maintenance. Production of the Y-20 large-transport aircraft, which entered service in July 2016, will likely provide relief on this front.[160]

Insufficient facilities abroad impose another serious constraint. To date, the most-common PLAAF overseas deployment remains the deployment of two to four large-transport aircrafts at a time for HA/DR, personnel recovery, or other nonwar missions. Security, logistics, and maintenance needs for larger groups of dissimilar aircraft impose serious constraints on the ability of the PLAAF to carry out expeditionary activities on a larger scale. In the future, the PLAAF will need to establish reliable access to overseas airfields if it hopes to operate missions of a broader variety and higher tempo abroad. The PLA Navy has already established such a precedent in arranging for a supply point in Djibouti in 2015.[161]

Through these deployments, small numbers of Chinese aircrews and technicians are learning to navigate abroad, manage issues of diplomatic access, and operate with greater autonomy. As the PLAAF gains experience through these activities, it is updating aspects of its approach to expeditionary operations, to include deployed communications, logistics, and maintenance. The acquisition of larger, more capable transport planes such as the Y-20, more experience in operating dissimilar aircraft, and greater access to foreign airfields will enable the PLAAF to better carry out its nonwar missions in Asia and around the world. Moreover, greater confidence in operating abroad will position the PLAAF to carry out a broader array of missions than it has hitherto performed. While the PLAAF's expeditionary deployments to date remain small and limited by Western standards, an increasing need to safeguard Chinese interests abroad suggests that the development of expeditionary capabilities will remain a priority for the PLAAF for years to come.

---

[160] "Chinese Large Freighter Plane Enters Military Service," Xinhua [新华] in English, July 7, 2016.

[161] "China's Logistic Hub in Djibouti to Stabilize Region, Protect Interests," *Global Times* in English, March 17, 2016.

Appendix. List of PLAAF International Deployments, 2002–2016

Table A.1 lists all 38 deployments mentioned in this report and featured in Figure 3.1. The data cover all observed deployments between January 2002 and October 2016.

Table A.1. Observed PLAAF Deployments, January 2002–October 2016

Deployment Type	Year	Exercise Name or Event Description	Location
Exercises-combat	2007	Peace Mission	Russia
Exercises-combat	2010	Anatolian Eagle	Turkey
Exercises-combat	2011	Shaheen I	Pakistan
Exercises-combat	2013	ADMM-Plus HADR/MM Ex	Brunei
Exercises-combat	2013	Peace Mission	Russia
Exercises-combat	2014	Shaheen III	Pakistan
Exercises-combat	2015	Peace and Friendship	Malaysia
Exercises-combat	2015	Falcon Strike	Thailand
Exercises-combat	2016	Shaheen V	Pakistan
Exercises-combat	2016	AM-HEx	Thailand
Exercises-airborne	2011	Divine Eagle/Condor	Belarus
Exercises-airborne	2011	Cooperation	Venezuela
Exercises-airborne	2013	Sharp Knife Airborne	Indonesia
Exercises-airborne	2014	Sharp Knife Airborne	Indonesia
Exercises-airborne	2015	Divine Eagle/Condor	Belarus
Exercises-airborne	2016	International Army Games "Airborne Platoon"	Russia
NEO	2011	NEO	Libya
HA/DR	2002	War relief	Afghanistan
HA/DR	2010	Winter storm relief	Mongolia
HA/DR	2010	Flood relief	Pakistan
HA/DR	2011	Flood relief	Pakistan

Deployment Type	Year	Exercise Name or Event Description	Location
HA/DR	2011	Flood relief	Thailand
HA/DR	2014	Water shortage relief	Maldives
HA/DR	2015	Flood relief	Malaysia
HA/DR	2015	Flood relief	Sri Lanka
HA/DR	2015	Earthquake relief	Nepal
HA/DR	2015	Flood relief	Myanmar
Personnel recovery	2004	Recover remains of deceased nationals	Afghanistan
Personnel recovery	2014	Search and rescue for Malaysian Airlines Flight 370	Malaysia
Personnel recovery	2014	Search and rescue for Malaysian Airlines Flight 370	Australia
Personnel recovery	2016	Recover remains of deceased nationals	South Korea
Air shows/competitions	2013	Bayi Team Moscow Air Show performance	Russia
Air shows/competitions	2014	Aviadarts 2014	Russia
Air shows/competitions	2015	Bayi Team Langkawi performance	Malaysia
Air shows/competitions	2015	Bayi Team Thailand stopover performance	Thailand
Air shows/competitions	2015	Aviadarts 2015	Russia
Air shows/competitions	2015	Bayi Team post–Falcon Strike performance	Thailand
Air shows/competitions	2016	Aviadarts 2016	Russia

Bibliography

"Air Force Aviation Force Flies Out into the World for Training, What News Has Been Transmitted?" ["空军航空兵飞出国门联训，传递了什么信息？"], *China Military Online* 《中国军网》, April 9, 2016. As of November 2, 2016: http://www.81.cn/syjdt/2016-04/09/content_6998435_3.htm

"Air Force Planes Transport Materiel to Nepal," *China Daily Online* in English, April 29, 2015. As of November 8, 2016: http://usa.chinadaily.com.cn/world/2015-04/29/content_20577519.htm

"Air Forces of China and Pakistan Benefit from Joint Training," *China Military Online* in English, September 17, 2015. As of April 25, 2017: http://english.chinamil.com.cn/news-channels/china-military-news/2015-09/17/content_6685561.htm

Allen, Kenneth, and Emma Kelly, "Assessing the PLAAF's Foreign Relations," *China Brief*, Vol. 12, No. 9, April 26, 2012. As of November 8, 2016: http://www.jamestown.org/single/?tx_ttnews[tt_news]=39304&no_cache=1#.Vv6eiWQrLJx

———, "China's Air Force Female Aviators: Sixty Years of Excellence (1952–2012)," *China Brief*, Vol. 12, No. 12, June 22, 2012. As of November 1, 2016: http://www.jamestown.org/single/?tx_ttnews%5Btt_news%5D=39532&no_cache=1#.VwqIXTYrLJw

Allen, Kenneth W., Glenn Krumel, and Jonathan D. Pollack, *China's Air Force Enters the 21st Century*, Santa Monica, Calif.: RAND Corporation, MR-580-AF, 1995. As of July 17, 2017: https://www.rand.org/pubs/monograph_reports/MR580.html

Allen, Kenneth, Phillip C. Saunders, and John Chen, "Chinese Military Diplomacy, 2003–2016: Trends and Implications," *China Strategic Perspectives* 11, Institute for National Strategic Studies, National Defense University, Washington, D.C., July 2017.

"Analysis: Similarities and Changes in Seven 'Peace Mission' War Games," Xinhua [新华] in English, September 17, 2010. As of November 2, 2016: http://news.xinhuanet.com/english2010/indepth/2010-09/17/c_13517547.htm

Andrew, Martin, "Power Politics: China, Russia, and Peace Mission 2005," *China Brief*, Vol. 5, No. 20, September 27, 2005. As of November 4, 2016: http://www.jamestown.org/single/?tx_ttnews%5Btt_news%5D=30909&no_cache=1#.VwK8BVK9aoM

"'Aviadarts 2014' Enters Air Race Stage," *China Military Online* in English, July 28, 2014. As of November 8, 2016: http://eng.chinamil.com.cn/news-channels/china-military-news/2014-07/28/content_6067048.htm

"'Aviadarts 2014' International Pilot Competition Ends," *China Military Online* in English, July 29, 2014. As of November 8, 2016: http://eng.chinamil.com.cn/news-channels/china-military-news/2014-07/29/content_6068830.htm

"'Aviadarts' International Contest Kicks Off," *China Military Online* in English, August 16, 2016. As of October 27, 2016: http://english.chinamil.com.cn/news-channels/china-military-news/2016-08/04/content_7190447.htm

"B-4018 PLAAF—China Air Force Boeing 737-33A," Planespotters.net, 2016. As of October 31, 2016: https://www.planespotters.net/photo/514719/b-4018-plaaf-china-air-force-boeing-737-33a

Begawan, Bandar Seri, "More Troops Arrive for Military Exercises," *Brunei Times Online* in English, June 10, 2013. As of November 4, 2016: http://www.everyday.com.kh/en/article/6246.html

Belousov, Yuriy [Юрий БЕЛОУСОВ], "In Merit There Is Honor" ["По заслугам и честь"], *Krasnaya Zvezda Online* 《Красная звезда》, August 22, 2013. As of November 2, 2016: http://redstar.ru/index.php/dudenko/item/11048-po-zaslugam-i-chest

Blasko, Dennis J., *The Chinese Army Today: Tradition and Transformation for the 21st Century*, 2nd ed., New York: Routledge, 2012.

Bo Xu [薄旭], "A Record-Breaking Major Evacuation" ["创纪录大撤离"], *World Knowledge* 《世界知识》, July 2011, pp. 47–52.

"Body of Chinese Peacekeeper Being Flown Back to China," *CRI English*, June 8, 2016. As of November 8, 2016: http://en.people.cn/n3/2016/0608/c90786-9069998.html

"C-130 Hercules," May 2014. As of January 3, 2017: http://www.af.mil/AboutUs/FactSheets/Display/tabid/224/Article/104517/c-130-hercules.aspx

"C-17 Globemaster III," October 2015. As of January 3, 2017: http://www.af.mil/AboutUs/FactSheets/Display/tabid/224/Article/104523/c-17-globemaster-iii.aspx

"C-5 A/B/C Galaxy and C-5M Super Galaxy," January 2014. As of January 3, 2017: http://www.af.mil/AboutUs/FactSheets/Display/tabid/224/Article/104492/c-5-abc-galaxy-and-c-5m-super-galaxy.aspx

Chase, Michael S., Kenneth W. Allen, and Benjamin S. Purser III, *Overview of People's Liberation Army Air Force "Elite Pilots,"* Santa Monica, Calif.: RAND Corporation, RR-1416-AF, 2016. As of July 17, 2017: http://www.rand.org/pubs/research_reports/RR1416.html

Che Mengtao [车孟涛] and Liu Yinghua [刘应华], "Air Force Flying Leopard Combat Aircraft High in the Sky Launch a Violent Assault" ["空军飞豹战机凌空突击"], Xinhua [新华] in English, August 29, 2016. As of October 27, 2016: http://www.81.cn/kj/2016-08/29/content_7235911.htm

"China Ahead in Asian States' Post–Cold War Battle," *Jane's Defence Weekly*, September 25, 1993.

"China Aircraft Arrives in ROK to Return Martyrs' Remains," *China Military Online* in English, March 29, 2016. As of November 8, 2016: http://eng.chinamil.com.cn/news-channels/china-military-news/2016-03/29/content_6982507.htm

"China and Malaysia Hold Peace and Friendship 2015 Exercise" ["中马举行 '和平友–2015' 联合军事演习"], August 27, 2015. As of November 1, 2016: http://news.mod.gov.cn/headlines/2015-08/27/content_4616108.htm

"China, Indonesia Complete Anti-Terror Exercise," *China Daily Online* in English, November 12, 2013. As of November 4, 2016: http://usa.chinadaily.com.cn/china/2013-11/12/content_17097501.htm

"China Initiates First Class Emergency Response for Quake Relief in Qinghai," Xinhua [新华] in English, April 14, 2010. As of July 18, 2017: http://www.chinaconsulate.org.nz/eng/gdxw/t681949.htm

"China, Malaysia Conclude First Joint Military Exercise," Xinhua [新华] in English, September 23, 2015. As of November 4, 2016: http://www.chinadaily.com.cn/world/2015-09/23/content_21956618.htm

"China Pledges $250m Flood Aid to Pakistan," *China Daily* in English, December 18, 2010. As of November 8, 2016: http://www.chinadaily.com.cn/china/2010wenindia/2010-12/18/content_11721909.htm

"China Ranks Second in 10 Team Competitions at International Military Games 2015," *China Military Online* in English, August 17, 2015. As of November 8, 2016: http://english.chinamil.com.cn/news-channels/2015-08/17/content_6635532.htm

"China Sends Record Military Personnel Numbers to Nepal," Xinhua [新华] in English, May 7, 2015. As of August 1, 2017: http://www.chinadaily.com.cn/china/2015-05/07/content_20652513.htm

"China to Participate in International Army Games 2016," *China Military Online* in English, July 1, 2016. As of November 8, 2016: http://english.chinamil.com.cn/news-channels/china-military-news/2016-07/01/content_7128942.htm

"China-Belarus First Airborne Joint Exercise Concludes Well" ["中国与白俄罗斯首次空降兵联合训练圆满结束"], Xinhua [新华], July 19, 2011. As of November 4, 2016: http://www.gov.cn/jrzg/2011-07/19/content_1909398.htm

"China-Malaysia Hold Peace and Friendship 2015 Joint Exercise" ["中马将举行 '和平友谊-2015' 联合军事演习"], Ministry of National Defense, August 27, 2015. As of November 4, 2016: http://news.mod.gov.cn/headlines/2015-08/27/content_4616108.htm

"China-Pakistan Air Forces Hold Shaheen V Joint Exercise in Pakistan" ["中巴空军在巴基斯坦举行 '雄鹰-V' 联合训练"], Ministry of National Defense, April 9, 2016. As of November 4, 2016: http://www.mod.gov.cn/action/2016-04/09/content_4648907.htm

"China-Thai Joint Air Force Drill: A Signal for More Open Chinese Air Force," *China Military Online* in English, November 17, 2015. As of November 4, 2016: http://english.chinamil.com.cn/news-channels/pla-daily-commentary/2015-11/17/content_6773537.htm

"China-Thailand 'Falcon Strike 2015' J-11s Encounter 'Gripen'" ["中泰 '鹰击－2015' 歼－11遇 '鹰狮'"], *CCTV News Online*《央视网》, November 12, 2015. As of November 1, 2016: http://news.cntv.cn/2015/11/12/VIDE1447340194183574.shtml

"China's First Military Air Ambulance Debuts," *China Military Online* in English, August 22, 2016. As of November 8, 2016: http://english.chinamil.com.cn/view/2016-08/22/content_7219353.htm

"China's JH-7 Wins Runner-Up of 'Aviadarts-2016' Competition in Russia," *China Military Online*, August 9, 2016. As of October 27, 2016:

http://english.chinamil.com.cn/news-channels/china-military-news/2016-08/09/content_7198410.htm

"China's Logistic Hub in Djibouti to Stabilize Region, Protect Interests," *Global Times* in English, March 17, 2016. As of November 8, 2016: http://english.chinamil.com.cn/news-channels/pla-daily-commentary/2016-03/17/content_6964486.htm

"Chinese Air Force Attends 'Aviadarts 2014' International Pilot Competition in Russia," *China Military Online* in English, July 21, 2014. As of November 8, 2016: http://eng.mod.gov.cn/DefenseNews/2014-07/21/content_4523563.htm

"Chinese Air Force Dispatches 3 Types, 5 Combat Aircraft to Participate in Sino-Russian Joint Military Exercise" ["中国空军出动3型5架战机参加中俄联合军演"], Central Government Web Portal [中央政府门户网站], August 24, 2015. As of August 7, 2017: http://www.gov.cn/xinwen/2015-08/24/content_2919059.htm

"Chinese Air Force Evacuates 751 from Libya," Xinhua [新华] in English, March 2, 2011. As of December 30, 2016: http://www.chinadaily.com.cn/china/2011-03/02/content_12103843.htm

"Chinese Air Force Sends Three Transport Aircraft to Deliver Relief Supplies for Thailand Flood," *People's Daily Online*《人民网》, October 22, 2011. As of May 2, 2016: http://military.people.com.cn/GB/172467/15984069.html

"Chinese Air Force to Fly Tomorrow to Afghanistan to Receive the Remains of Victims" ["中国空军专机明日赴阿富汗接遇难者遗体"], Southcn.com [南方网], June 13, 2004. As of November 8, 2016: http://www.southcn.com/news/international/pic/200406130206.htm

"Chinese Air Force Visits Belarus to Participate in Joint Counter-Terrorism Exercise (Pictures)" ["中国空降兵赴白俄罗斯参加联合反恐训练(图)"], *China News Online*《中国新闻网》, June 15, 2015. As of November 4, 2016: http://www.chinanews.com/mil/2015/06-15/7344467.shtml

"Chinese Air Force's First Assistance Aircraft Arrives in Nepal Earthquake Disaster Area" ["中国空军首架救援飞机飞抵尼泊尔地震灾区"], *China Military Online*《中国军网》, April 28, 2017. As of August 7, 2017: http://www.81.cn/kjzj/2015-04/28/content_6954755.htm

"Chinese Airmen Hone Skills in International Competition," Xinhua [新华] in English, August 28, 2016. As of November 8, 2016: http://www.chinadaily.com.cn/china/2016-08/28/content_26619315.htm

“Chinese Ambassador to Djibouti States Yemen Evacuation Work Proceeded Smoothly” [“中国驻吉布提大使说也门撤侨工作顺利进行”], *People's Daily*《人民网》, April 1, 2015. As of November 8, 2016:
http://world.people.com.cn/n/2015/0401/c157278-26783520.html

“Chinese Defense Attachés Participate in All-Out Task to Evacuate Chinese Personnel in Libya” [“中国武官全力参与在利比亚中方人员撤离工作”], Xinhua [新华], February 28, 2011. As of December 30, 2016:
http://www.chinanews.com/gn/2011/02-28/2873609.shtml

“Chinese Government's Third Batch of Relief Supplies to Thailand Arrives in Bangkok” [“中国政府向泰国提供的第三批援助物资运抵曼谷”], Central Government Web Portal [中央政府门户网站], October 22, 2011. As of July 18, 2017:
http://www.gov.cn/jrzg/2011-10/22/content_1975933.htm

“Chinese Il-76 to Conduct Heavy-Equipment Air Dropping,” *China Military Online* in English, August 22, 2014. As of November 4, 2016:
http://eng.mod.gov.cn/DefenseNews/2014-08/22/content_4531503.htm

“Chinese Large Freighter Plane Enters Military Service,” Xinhua [新华] in English, July 7, 2016. As of November 8, 2016:
http://english.chinamil.com.cn/news-channels/china-military-news/2016-07/07/content_7139555.htm

“Chinese Military Delivers Aid to Malaysia,” Xinhua [新华] in English, January 12, 2015. As of November 8, 2016:
http://www.globaltimes.cn/content/901478.shtml

“Chinese Military Team Starts Working in Gao of Mali,” *China Military Online* in English, June 6, 2016. As of November 8, 2016:
http://english.chinamil.com.cn/news-channels/china-military-news/2016-06/06/content_7089022.htm

“Chinese Military Transport Planes Carry Ammunitions to Russia for Int'l Games,” *China Military Online* in English, July 30, 2015. As of November 8, 2016:
http://english.chinamil.com.cn/news-channels/china-military-news/2015-07/30/content_6606037.htm

“Chinese Paratroopers Prepare for Int'l Army Games 2016,” *China Military Online* in English, July 27, 2016. As of November 8, 2016:
http://english.chinamil.com.cn/news-channels/china-military-news/2016-07/27/content_7176979.htm

"Chinese Peacekeeper's Body Brought Home from Mali," Xinhua [新华] in English, June 9, 2016. As of November 8, 2016: http://english.chinamil.com.cn/news-channels/china-military-news/2016-06/09/content_7093897.htm

"Chinese Troops Arrive in Thailand to Participate in the Second ASEAN Defense Ministers' Meeting Plus Humanitarian Assistance Disaster Relief and Military Medicine Joint Drill" ["中国军队抵达泰国参加第二次东盟防长扩大会人道主义援助救灾与军事医学联合演练"], Ministry of National Defense, September 1, 2016. As of November 1, 2016: http://www.mod.gov.cn/topnews/2016-09/01/content_4722991.htm

"Direct Hit China-Thai Air Forces' First Joint Flight Exhibition" ["直击中泰空军首次联合飞行表演"], *China Daily Online*《中国日报中文网》, November 30, 2015. As of November 8, 2016: http://www.chinadaily.com.cn/hqcj/zxqxb/2015-11-30/content_14367959.html

Directory of PRC Military Personalities, Washington, D.C.: Defense Intelligence Agency, 2016.

Dong Ruifeng [董瑞丰] and Guo Hongbo [郭洪波], "Sharp Weapons of a Big Power: Visit to an Air Force Strategic Transport Aircraft Unit," ["大国利器: 探访空军战略运输机部队"] *Outlook*《了望》, No. 34, August 20, 2012, pp. 48–49.

Dong Wenxian [董文先], "The Expansion of National Strategic Space Calls for a Strategic Air Force" ["国家战略空间扩展呼唤战略空军"], *Air Force News* 《空军报》, February 2, 2008, p. 2.

Erickson, Andrew, "China's Modernization of Its Air and Naval Capabilities," in Ashley Tellis and Travis Tanner, eds., *Strategic Asia 2012–13: China's Military Challenge*, Washington, D.C.: National Bureau of Asian Research, 2012, p. 76.

"Facts and Figures: Chinese Air Force Observes 65th Anniversary," Xinhua [新华] in English, November 4, 2014. As of November 8, 2016: http://news.xinhuanet.com/english/china/2014-11/04/c_127178409.htm

Fan Haisong [范海松] and Ding Yanli [丁艳丽], "Capability of a Guangzhou MR Transport Aviation Division's Long Range Airlift and Long Range Mobility Operations Ascends" ["广州运输航空兵某师远程空运和远程机动作战能力跃升"], *Air Force News*《空军报》, March 26, 2005, p. 1.

Garafola, Cristina L., "The Evolution of the PLA Air Force's Mission, Roles and Requirements," in Joe McReynolds, ed., *China's Evolving Military Strategy*, Washington, D.C.: The Jamestown Foundation, 2016, pp. 75–100.

"'Golden Darts' Comes to an End; Chinese Air Force Participants Seize Second Place" ["'航空飞镖'闭幕 中国空军参赛队夺第二名"], *China Military Online* 《中国军网》, August 10, 2016. As of November 8, 2016: http://www.81.cn/kj/2016-08/10/content_7199796.htm

Grevatt, Jon, "China's Il-76/Il-78 Order from Russia Faces Setback," *Jane's Defence Industry*, December 18, 2008.

Guo Hongbo [郭洪波], "Three Chinese Air Force Transport Aircraft Carry Relief Provisions to Pakistan" ["中国空军 3 架运输机向巴基斯坦运送救灾物资"], *China News Service* 《中国新闻网》, August 4, 2010. As of November 8, 2016: http://www.chinanews.com/gn/2010/08-04/2445873.shtml

Hartnett, Daniel, "The New Historic Missions: Reflections on Hu Jintao's Military Legacy," in Roy Kamphausen, David Lai, and Travis Tanner, eds., *Assessing the PLA in the Hu Jintao Era*, Carlisle, Pa.: Army War College Press, 2014, pp. 31–80.

"Heeding Party Commands, Serving the People, and Fighting Bravely and Skillfully" ["听党指挥 服务人民 英雄善战"] series, CCTV-7 "*Military Report*" 《"军事报道"》 program, December 7, 2006.

Hu Shanmin [胡善敏], Kou Lihua [寇利华], Li Wei [李伟], Xia Chenghao [夏程浩], and Li Xiulin [李秀林], "Four Subjects a Day Set Record of Highest Density of Drills" ["一天联演四课目 演习密度最高"], CCTV News "Focus On" program, August 28, 2015.

Hu Shanmin [胡善敏], Xu Miaobo, Li Wei [李伟], Xia Chenghao [夏程浩], and Cheng Pengcheng, "China and Russia Conclude 'Joint Sea 2015 (II)' Exercise," CCTV News "Focus On" program, August 28, 2015.

"Il-76MD Transport Plane Delivered to China," *The Moscow Times*, January 24, 2013. As of November 1, 2016: https://themoscowtimes.com/articles/il-76md-transport-plane-delivered-to-china-20953

"Il-76s in a Hurry: Taking Stock of Major Operations by Chinese Air Force Il-76 Transport Aircraft" ["伊尔很忙: 盘点中国空军伊尔-76 运输机重大行动"], Sina Photographs, December 9, 2014. As of November 8, 2016: http://slide.mil.news.sina.com.cn/k/slide_8_62085_33151.html#p=1

"In Brief Chinese AEW-A Correction, *Jane's Defence Weekly*, September 7, 1997.

"In Erdogan's Skies" ["Nei cieli di Erdogan"], *Milan Il Foglio* [*The Milan Paper*], October 14, 2010, p. 1.

"Injured Peacekeepers Return Home from South Sudan," Xinhua [新华] in English, July 17, 2016. As of November 8, 2016: http://eng.mod.gov.cn/HomePicture/2016-07/17/content_4695145.htm

International Institute for Strategic Studies, *The Military Balance 2017*, London: Routledge, 2017.

"JH-7 Fighter Bombers Participate in International Games in Russia," *China Military Online*, August 4, 2016. As of October 27, 2016: http://english.chinamil.com.cn/news-channels/china-military-news/2016-08/04/content_7191082.htm

Jiang, Steven, "China's Military Gets Boost with Huge New Transport Plane," *CNN*, July 7, 2016. As of November 8, 2016: http://www.cnn.com/2016/07/07/asia/china-jumbo-freighter

"Jinan Military Region Requisitions Civilian Aircraft for Long-Distance Transfer of Exercise Troops" ["济南军区参演部队征用民航飞机远程投送兵力"], Xinhua [新华], August 17, 2009. As of November 9, 2016: http://news.xinhuanet.com/mil/2009-08/17/content_11898840.htm

"The Joint Drill Between China and Venezuela Is a New Trial Exercise" ["中委两军组织联合训练，这是我们一种新的尝试"], film clip with Major General Zheng Yuanlin, CCTV News Content, November 9, 2011.

Le Tian, "Peace Mission Exercises Get Under Way," *China Daily Online* in English, August 7, 2007, p. 3. As of November 1, 2016: http://www.chinadaily.com.cn/cndy/2007-08/07/content_5448835.htm

Li Daguang [李大光], "PLA 'First-Time' Achievements in 'Peace Mission 2007' Exercise" ["解放軍在 '和平使命－2007' 演習中的 '第一次'"], *Wen Wei Po*《文匯網》, August 24, 2007.

Li Guowen [李国文], Wu Dechao [吴德超], Zhu Bin [朱斌], Zheng Wei [郑蔚], and Qian Bei [钱蓓], "Decoding the People's Air Force's Core Military Capabilities" ["解读人民空军核心军事能力"], *Wen Hui News*《文汇报》, November 2, 2009.

Li Qiang [李强] and Yin Jiahua [尹家骅], "Air Force Logistics Department Organizes Special Topic Seminar on Accelerating the Building of Comprehensive Support Capabilities" ["空军后勤部组织专题研讨 加快推进部队高原常态化驻训综合保障能力建设"], *Air Force News*《空军报》, June 7, 2012, p. 1.

Liu Gang [刘纲] and Yu Yi [余毅], "Establish Scientific Mechanisms, Oversee Each Phase: Airborne Air Transport Regiment Training and Tasks Progress Steadily" ["建立科学机制 严把每个环节: 空降某航运团训练和任务同向稳步推进"], *Air Force News*《空军报》, March 26, 2015, p. 1.

"Love Without Limits, Surge of Compatriot Feelings! Chinese Government's Myanmar Disaster Relief Assistance Activities; Temporary Housing Handover Ceremony Held in Yangon Harbor" ["爱无界 胞波情！ 中国政府援助缅甸活动板房交接仪式在仰光港举行"], *Myanmar Golden Phoenix*《金凤凰》, March 11, 2016. As of December 30, 2016: http://www.mmgpmedia.com/general-news/12756-%E7%88%B1%E6%97%A0%E7%95%8C-%E8%83%9E%E6%B3%A2%E6%83%85%EF%BC%81-%E4%B8%AD%E5%9B%BD%E6%94%BF%E5%BA%9C%E6%8F%B4%E5%8A%A9%E7%BC%85%E7%94%B8%E6%B4%BB%E5%8A%A8%E6%9D%BF%E6%88%BF%E4%BA%A4%E6%8E%A5%E4%BB%AA%E5%BC%8F%E5%9C%A8%E4%BB%B0%E5%85%89%E6%B8%AF%E4%B8%BE%E8%A1%8C

Luo Zheng [罗铮] and Zhu Hongliang [朱鸿亮], "Chinese Military Working Group Takes Military Aircraft and Rushes to South Sudan" ["中国军队工作组乘军机赶赴南苏丹"], *China Military Online*《中国军网》, July 13, 2016. As of November 8, 2016: http://www.81.cn/jmywyl/2016-07/13/content_7153316.htm

Mahadzir, Dzirhan, "China, Malaysia Hold First Bilateral Field Drills," *Jane's Defence Weekly*, September 18, 2015. As of November 1, 2016: https://janes.ihs.com/Janes/Display/1752781

"Malaysian Airlines MH370 Search Advances: Chinese Air Force Il-76 and Y-8 Deploy to Malaysia," ["马航 MH370 搜救进展：中国空军伊尔 76 运 8 飞赴大马"], *New Look*《新观察》, March 21, 2014. As of November 8, 2016: http://www.newjunshi.net/news/junqingguancha/53559_2.html

McDermott, Roger, "The Rising Dragon: Peace Mission 2007," Jamestown Foundation occasional paper series, October 2007. As of November 4, 2016: https://jamestown.org/report/the-rising-dragon-sco-peace-mission-2007

"Military Aircraft Bring Back 287 Chinese from Libya," Xinhua [新华] in English, March 5, 2011. As of December 30, 2016: http://en.people.cn/90001/90776/90883/7308966.html

Military Training Department of the General Staff of the Chinese People's Liberation Army, *Research into the Kosovo War*, Beijing: Liberation Army Publishing House [解放军出版社], 2000.

"Mission Action–2013 Cross-Regional Maneuver Exercise: Military-Civilian Air Transport Forces Jointly Deliver Lightly-Equipped Units" ["'使命行动–2013' 跨区机动演习: 军地空中运力联合投送轻装部队"], CCTV-13 News "Live News" [新闻直播间] program, September 15, 2013.

"Nine Chinese Aircraft Finish Ebola Rescue Task," Xinhua [新华], November 19, 2014. As of November 1, 2016: http://english.chinamil.com.cn/news-channels/2014-11/19/content_6232089.htm

"Our Country Deploys Il-76s to Pakistan to Deliver Relief Supplies" ["我空军出动伊尔 76 运输机向巴基斯坦运送救援物资"], *China Online*《中国网》, September 23, 2011. As of November 8, 2016: http://www.china.com.cn/military/txt/2011-09/23/content_23475779.htm

"Our Military Expands Civil-Military Fusion to News Roads of Aviation Emergency Response Assistance" ["我军拓展军民融合航空应急救援新路"], *PLA Daily*《解放军报》, August 23, 2016. As of November 8, 2016: http://military.people.com.cn/n1/2016/0823/c1011-28657227.html

"Our Military Working Group Goes to Mali to Launch Efforts Followed Ambushed Peacekeeping Personnel Casualties; Conduct Rescue and Medical Treatment Work" ["我军工作组赴马里开展遇袭伤亡维和人员善后和救治工作"], *China Military Online*《中国军网》, June 2, 2016. As of November 8, 2016: http://www.81.cn/sydbt/2016-06/02/content_7084129.htm

"Pak-China Air Exercise Shaheen V Begins," *Radio Pakistan* in English, April 15, 2016. As of November 4, 2016: http://www.radio.gov.pk/15-Apr-2016/pak-china-joint-air-exercise-shaheen-v-begins

Patel, Nirav, "Chinese Disaster Relief Operations: Identifying Critical Capability Gaps," *Joint Forces Quarterly*, No. 52, first quarter 2009, pp. 111–117.

"'Peace Mission 2007' to Test Remote Mobile Ability of Chinese Forces," *PLA Daily*《解放军报》. As of November 1, 2016: http://en.people.cn/90002/91620/91644/6225682.html

People's Liberation Army, *People's Liberation Army Military Terminology* 《中国人民解放军军语》, Beijing: Military Science Press, 2011.

Pettyjohn, Stacie L., *U.S. Global Defense Posture: 1783–2011*, Santa Monica, Calif.: RAND Corporation, MG-1244-AF, 2012. As of July 18, 2017: https://www.rand.org/pubs/monographs/MG1244.html

PLA Academy of Military Science Military Strategy Research Department, ed., *The Science of Military Strategy* 《战略学》, Beijing: Military Science Press, 2013, pp. 222–224.

"PLA Air Force J-11s Visit Thailand to Challenge Gripen Fighters—Reports from the Scene" ["解放军多架歼 11 赴泰叫阵鹰狮战机 现场曝光"], *Global Times* 《环球时报》, November 19, 2015. As of November 4, 2016: http://mil.sohu.com/20151119/n427023105.shtml

"PLA Airborne Commando Returns from Anti-Terrorism Drill," *PLA Daily Online* in English, November 23, 2011. As of November 8, 2016: http://thechinatimes.com/online/2011/11/1947.html

"PLA Details Chinese Military Operations Other than War Since 2008," *PLA Daily* in English, September 6, 2011. As of November 8, 2016: http://www.researchinchina.com/Htmls/News/201109/22169.html

"PLAAF Aerobatic Team Arrives in Malaysia for Stunt Shows," *China Military Online* in English, March 12, 2015. As of November 8, 2016: http://eng.mod.gov.cn/DefenseNews/2015-03/12/content_4574587.htm

"PLAAF Combat Group Holds First Flight Training" ["中方空军战斗群首次飞行训练"], CCTV-13 News [CCTV-13 新闻] program, August 7, 2013.

"PLAAF Overcomes Disadvantages to Commit Full Efforts to Searching Missing Plane," *China Military Online* in English, March 28, 2014. As of November 8, 2016: http://eng.chinamil.com.cn/special-reports/2014-03/28/content_5834154.htm

"PLAAF Su-27 Fighters Refueled in Iran While Traveling to Turkey for Exercise" ["中国空军苏-27 战机赴土耳其途中曾在伊朗加油"], *Dongfang Online* 《东方网》, October 11, 2012. As of November 4, 2016: http://mil.news.sina.com.cn/2010-10-12/0946614116.html

"Plane Carrying Body of Deceased Chinese Peacekeeper Has Arrived in Changchun," CCTV.com, June 13, 2016. As of November 8, 2016: http://english.chinamil.com.cn/2006radioscom/2016-06/13/content_7098946.htm

Qiu Renjie [邱仁傑], "Disaster Relief for an Entire People; Malaysia-China Live Troop Exercise Preparations" ["全民賑災 馬中實兵演習前熱身"], *China Press* 《中國報》, January 13, 2015.

Qu Yantao [曲延涛], Yang Qinggang [杨清刚], and Xu Xiaolong [徐小龙], "Peace and Friendship 2015 China Malaysia Maritime Drill," *China Military Online* 《中国军网》, September 20, 2015. As of November 4, 2016: http://jz.chinamil.com.cn/n2014/tp/content_6689744.htm

"Record of Chinese Air Force's Search for MH370," *China Military Online* in English, April 11, 2014. As of October 31, 2016:
http://eng.mod.gov.cn/Database/MOOTW/2014-04/11/content_4503333.htm

"Remains of Chinese Peacekeepers Brought Home," *China Military Online* in English, July 20, 2016. As of November 8, 2016:
http://eng.mod.gov.cn/DefenseNews/2016-07/20/content_4697472.htm

Rupprecht, Andreas, and Tom Cooper, *Modern Chinese Warplanes: Combat Aircraft and Units of the Chinese Air Force and Naval Aviation*, Houston: Harpia Publishing, 2012.

"Russia and China Agree Sale of Transport Aircraft and Fuel Dispensers," *Jane's Defence Industry*, September 9, 2005.

"SCO Leaders Observe Joint Anti-Terror Drill," Xinhua [新华] in English, August 17, 2007. As of November 1, 2016:
http://english.sina.com/1/2007/0817/122068.html

"Second AM-HEx 2016 Joint Exercise Underway in Thailand," *China Military Online* in English, September 8, 2016. As of November 4, 2016:
http://english.chinamil.com.cn/view/2016-09/08/content_7247931.htm

"'Sharp Knife Airborne–2014' Exercise Concludes," *China Military Online* in English, November 5, 2014. As of November 4, 2016:
http://english.chinamil.com.cn/news-channels/photo-reports/2014-11/05/content_6212872.htm

Sina.com, "Il-76 Transport Aircraft Large-Scale Activity," December 9, 2014. As of July 19, 2017:
http://slide.mil.news.sina.com.cn/k/slide_8_62085_33151.html#p=6.

Sun Xingwei [孙兴维] and Li Yunpeng [李云鹏], "First Y-9 Type Transport Aircraft Officially Enters Services with Army Aviation Forces" ["首架运-9 型运输机正式列装陆军航空兵部队"], *China Army Online*《中国陆军网》, December 23, 2016. As of April 27, 2017:
http://www.mod.gov.cn/power/2016-12/23/content_4767722.htm

Tan Jie, "PLA Air Force Transporters Evacuate Compatriots from Libya," *China Military Online* in English, March 2, 2011.

Tanakasempipat, Patpicha, and Jutarat Skulpichetrat, "China, Thailand Joint Air Force Exercise Highlights Warming Ties," Reuters, November 24, 2015. As of November 8, 2016:
http://www.reuters.com/article/us-china-thailand-military-idUSKBN0TD0B120151124

"Ten Highlights of China's Military Diplomacy in 2013," *PLA Daily* in English, December 23, 2013.

"Three Large Transport Aircraft of the Chinese Air Force Congregate at the Shijiazhuang Zhending International Airport" ["中国空军 3 架大型运输机在石家庄正定国际机场集结"], Sina Photographs [图片], February 8, 2010. As of November 8, 2016: http://slide.mil.news.sina.com.cn/slide_8_193_2808.html#p=1

"Turkey, China Conduct Joint Air Maneuvers," *Istanbul Today's Zaman Online* in English, September 30, 2010.

U.S. Air Force, *Global Vigilance, Global Reach, Global Power for America*, August 2013, p. 7. As of November 8, 2016: http://www.af.mil/Portals/1/images/airpower/GV_GR_GP_300DPI.pdf

U.S. Air Force Historical Studies Office, "Evolution of the Department of the Air Force," May 4, 2011. As of October 31, 2017: http://www.afhso.af.mil/topics/factsheets/factsheet.asp?id=15236

U.S. Department of Defense, *DoD Dictionary of Military and Associated Terms*, July 2017, p. 85. As of July 23, 2017: http://www.dtic.mil/doctrine/new_pubs/dictionary.pdf

Wang Ran [王冉] and Tian Wei [田炜], "Tenth Air Force CCP Committee Holds Its 11th Plenary (Enlarged) Meeting in Beijing," ["空军党委十届十一次全体（扩大）会议在京召开"] *Air Force News* 《空军报》, December 30, 2008, p. 1.

Wang Xiaoyu [王晓玉], Yan Qian [闫倩], and Zhou Limin [周立民], "Air Force Combat Aircraft Participating in 'International Military Competition–2016' Arrive in Russia" ["空军参加 '国际军事比赛–2016' 战机抵俄"], CCTV-7 "Military Report" [军事报道] program, July 26, 2016.

Wang Yongming, Liu Xiaoli, and Xiao Yunhua, *Research into the Iraq War*, Beijing: Liberation Army Publishing House [解放军出版社], March 2003.

Wolf, Jim, "China Mounts Exercise with Turkey, U.S. Says," Reuters, October 8, 2010. As of November 4, 2016: http://www.reuters.com/article/us-china-turkey-usa-idUSTRE6975HC20101008

Xu Xieqing [徐叶青] and Ren Xu [任旭], "Sino-Russian 'Joint Sea–2015 (II)' Live Troop Exercise Fires First Shots" ["中俄'海上联合-2015（Ⅱ）'实兵演习打响"], *China Military Online*《中国军网》, August 24, 2015. As of August 7, 2017: http://www.81.cn/2015918/2015-08/24/content_6686419.htm

Yang Tiehu [杨铁虎] "Air Force Confirms: 'PLAAF Defeated by Turkey in Air Combat Exercise' Is Simply a Rumor" ["空军证实: '中国空军空战惨败土军' 纯属谣言"] *People's*

Daily 《人民日报》, October 15, 2010. As of November 4, 2016: http://military.people.com.cn/GB/172467/12966524.html

Yang Yongsheng [杨永生], Zhu Zhanghu [朱章虎], and Zhao Lingyu [赵凌芋], "Training that Builds 'Iron Wings'" ["练就'铁翅膀'"], *Air Force News* 《空军报》, August 21, 2012, p. 1.

Zambelis, Chris, "Sino-Turkish Strategic Partnership: Implications of Anatolian Eagle 2010," *China Brief*, Vol. 11, No. 1, January 14, 2011. As of November 4, 2016: http://www.jamestown.org/single/?tx_ttnews%5Btt_news%5D=37369#.Vv6gF2QrLJw

Zhang Changsheng [张长胜], Ma Junhong [马军鸿], and Li Benqian [李本钱], "Air Force Signals Regiment Participates in Joint Military Exercise Peace Mission 2007" ["空军某通信团参加 '和平使命–2007' 联合军演纪事"], *Air Force News* 《空军报》 September 18, 2007, p. 1.

Zhang Haiping, Zhang Feng, Zhang Zhe, Ouyang Dongmei, and Zhang Tao, "Peace Mission 2013: China-Russia Joint Anti-Terrorism Exercise" *China Military Online* in English, August 15, 2013. As of November 2, 2016: http://english.chinamil.com.cn/special-reports/node_60757.htm

Zhang Jinyu and Shen Jinke, "PLA Air Force Transporters Bring Home Chinese Evacuees from Libya," *China Military Online* in English, March 4, 2011.

Zhang Zhongkai [张钟凯] and Wu Dengfeng [吴登峰], "Military Experts Explain Highlights of Sino-Russian 'Joint Sea–2015 (II)' Exercise in Detail" ["军事专家详解中俄 '海上联合–2015 (II)' 演习亮点"], Xinhua [新华网], August 25, 2015. As of October 31, 2016: http://gb.cri.cn/48519/2015/08/25/6351s5078424.htm

Zhang Ziyang [张子扬], "Chinese Armed Forces Ship Ten Million RMB Worth of Aid Materials to Sri Lanka" ["中国军队启运 1000 万元援助斯里兰卡物资"], *China News Service* 《中国新闻网》, January 25, 2011. As of November 8, 2016: http://www.chinanews.com/gn/2011/01-25/2808767.shtml

Zhao Zongqi [赵宗岐], "Put the Construction of Strategic Delivery Capability in an Important Position" ["把战略投送能力建设摆到重要位置"], *PLA Daily* 《解放军报》, April 23, 2009, p. 10.

Zheng Jinran, "Peacekeeper's Body Returns for Burial," *China Daily* in English, June 10, 2016. As of November 8, 2016: http://wap.chinadaily.com.cn/2016-06/10/content_25663352.htm